Radiative Properties of Semiconductors

Radiative Properties of Semiconductors

N M Ravindra
New Jersey Institute of Technology Newark, New Jersey USA

Sita Rajyalaxmi Marthi
New Jersey Institute of Technology Newark, New Jersey USA

Asahel Bañobre
New Jersey Institute of Technology Newark, New Jersey USA

Morgan & Claypool Publishers

Rights & Permissions
To obtain permission to re-use copyrighted material from Morgan & Claypool Publishers, please contact info@morganclaypool.com.

ISBN   978-1-6817-4112-3 (ebook)
ISBN   978-1-6817-4048-5 (print)
ISBN   978-1-6817-4240-3 (mobi)

DOI   10.1088/978-1-6817-4112-3

Version: 20170801

IOP Concise Physics
ISSN 2053-2571 (online)
ISSN 2054-7307 (print)

A Morgan & Claypool publication as part of IOP Concise Physics
Published by Morgan & Claypool Publishers, 40 Oak Drive, San Rafael, CA, 94903 USA

IOP Publishing, Temple Circus, Temple Way, Bristol BS1 6HG, UK

This book is dedicated to the giants of science whose shoulders we stand up on and who continue to be the immortal guides to the progress of science and technology and to the creation of a better world.

Contents

Preface

Optical properties, particularly in the infrared range of wavelengths, continue to be of enormous interest to both material scientists and device engineers. The need for the development of standards for data of optical properties in the infrared range of wavelengths is very timely considering the on-going transition of nano-technology from fundamental R&D to manufacturing. The recent progress in two-dimensional materials is an example of this evolution in materials science and engineering.

Radiative properties play a critical role in the processing, process control and manufacturing of semiconductor materials, devices, circuits and systems. The design and implementation of real-time, non-contact process monitoring and control methods in manufacturing, such as multi-wavelength imaging pyrometry, spectro-scopic ellipsometry and reflectometry, require the knowledge of the radiative properties of materials.

The design and manufacturing of sensors, imagers, waveguides, filters, anti-reflection coatings and lenses, operating in the infrared range of wavelengths, requires a reliable database of the radiative properties of materials.

This book reviews the optical properties of various semiconductors in the infrared range of wavelengths. Some fundamental and experimental studies of the radiative properties of semiconductors are presented. Previous studies, potential applications and future developments are outlined.

In chapter 1, an introduction to the radiative properties is presented. A brief overview of the optical and thermal properties is presented in chapter 2. Examples of the instrumentation for the measurements of the radiative properties are described in chapter 3. In chapters 4–13, case studies of the radiative properties of several semiconductors are elucidated. The modeling and applications of these properties are explained in chapters 14 and 15, respectively. In chapter 16, examples of the global infrastructure for these measurements are illustrated.

Foreword

This year marks the 25th anniversary [1] since the Defense Advanced Research Projects Agency (DARPA) and the US Air Force Wright Laboratory contracted with Texas Instruments (TI) [2, 3] to develop a next generation flexible semiconductor wafer fabrication system called the Microelectronics Manufacturing Science & Technology (MMST). The $100 million, MMST Program, was a 5-year R&D effort, initiated under the leadership of Dr Arati Prabhakar (Director, DARPA, July 2012–January 2017), to develop a wide range of new technologies for advanced semiconductor manufacturing. The major highlight of the program was the demonstration, in 1993, of sub-3-day cycle time for manufacturing 350-nm CMOS integrated circuits. This was mainly enabled by the development of 100% single-wafer processing.

Around the same time, work had begun on the development of a Multi-Wavelength Imaging Pyrometer (M-WIP) (figure 1), at the New Jersey Institute of Technology (NJIT), under the sponsorship of the US Air Force Wright Laboratory and DARPA (then ARPA)—Microelectronics Technology Office.

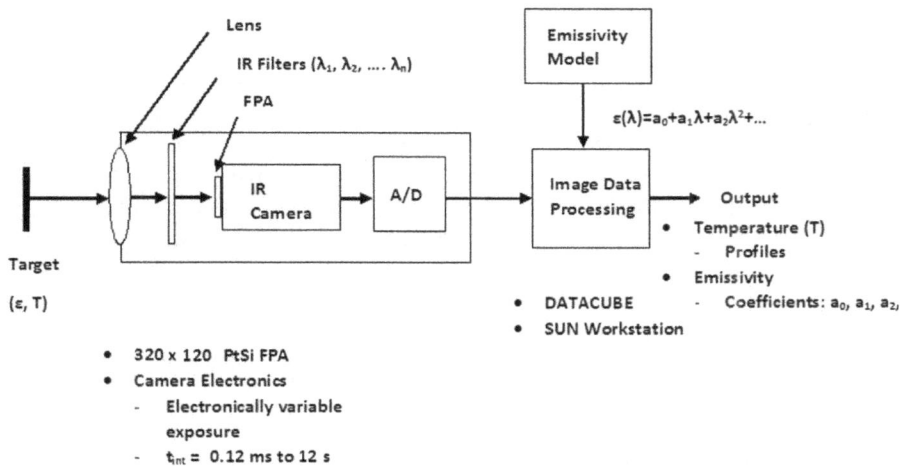

Figure 1. Multi-Wavelength Imaging Pyrometer [4].

The main components for the development of M-WIP for applications in remote sensing of temperature profiles of targets with unknown emissivity were as follows:
- Lens or array of micro lenses
- Infrared filters
- IR Camera/Focal Plane Array (FPA)
- Read Out Integrated Circuits/Electronics
- Image Data Processing Unit
- Control System Software
- Emissivity Model.

Under the sponsorship of the US Department of Defense/Defense University Research Instrumentation Program (DURIP), NJIT acquired a Spectral Emissometer (SE). By utilizing an in-line Fourier transform infrared (FTIR) spectrometer, SE was designed specifically to facilitate simultaneous measurements of surface spectral emittance and temperature by using optical techniques over the near and mid-IR spectral range and temperatures ranging from 300 K to 2000 K [5]. This non-contact, real-time technique enabled the measurements of the radiative properties of materials as a function of temperature and wavelength. Advanced Fuel Research (James Markham) [6] was responsible for the development of the Spectral Emissometer.

The US Department of Energy/National Renewable Energy Laboratory (NREL), Sematech (Don Lindholm, Arun Nanda, T Speranza) and several Semiconductor Research Corporation (William T Lynch, William Holton) member companies joined in this effort.

I acknowledge with thanks the fruitful discussions, over the years, with Advanced Fuel Research (James Markham); BTL Fellows Inc. (Martin P Lepselter); Cavendish Labs, Cambridge University (Paul Timans, K U Ahmed); City University of New York (Jacob Trevino); David Sarnoff Research Center (Bawa Singh, Michael Leahy, David Richman, Jon Thomas); D2IR Inc. (Peter Kaufman); Georgia Tech (Zhuomin Zhang); Harmon Sensors (Alex Stein); Massachusetts Institute of Technology (Jeff Hebb); National Institute of Standards and Technology (NIST—the late David Dewitt, Raju Datla, Benjamin K Tsai, Leonard M Hanssen); National Renewable Energy Laboratory (Bhushan Sopori); Online Technologies (Peter Solomon); Rochester Institute of Technology (Lynn Fuller and Santosh Kurinec); Stanford University (Krishna Saraswat, Mehrdad Moslehi); US Army at Fort Monmouth (Richard Lareau); Wright Patterson Air Force Base; several companies including Advanced Micro Devices; Applied Materials (Bruce Adams); AST-Steag (Jeff Gelpey); CVC (Mehrdad Moslehi); Dainippon Screen; Emcore; Epitax (Greg Olson, Krishna Linga); IBM; Intel; Lucent Technologies (Anthony T Fiory); Luxtron (Charles Schittinger); Philips (Fred Roozeboom); Rockwell International (Krishna Bajaj, Stuart Irvine); Sensarray; Texas Instruments (Pallab Chatterjee) and Vortek.

The participation and support of the late Chancellor Gary Thomas, late Raymond Balcerack (DARPA), late Walter F Kosonocky, late Eugene Gordon, late Constantin Manikopoulos, John C Hensel, Edwin Hou and Bernard Friedland is acknowledged with thanks. The team at NJIT consisted of a large number of scientists including Sergej Belikov, Oktay Gokce, Viphul Patel and Feiming Tong. Some of the research summarized in this book is based on the results that were obtained by an inter-disciplinary team of graduate students from the College of Computing Sciences, College of Engineering and the College of Science and Liberal Arts at NJIT. This includes Sufian Abedrabbo (U. Jordan), Wei Chen, Michael Kaplinsky (Arecont Vision), Chiranjivi Lamsal (Fort Peck Community College and U. Montana), Jun Li (Tsinghua), Nathaniel Mcaffrey (New Hampshire), Sarang Muley (National Standard), Yi Zhang and many others.

The late 1980s witnessed a flurry of activities on research relating to single wafer processing—with a goal to reduce the thermal budget (temperature time product), reduce process-induced contamination and increase process yield. Several equipment manufacturers focused on rapid thermal processing (RTP) as the approach to single-wafer manufacturing. These manufacturers included AET Thermal, AG Associates, Applied Materials, AST Steag, BCT Spectrum, CVC, Dainippon Screen, Eaton, G Squared Semiconductor Corp, General Electric, Jiplec, LEISK, Matson, Peak Systems, Process Products, Tamarack, TEL/Thermco, Texas Instruments, Varian and Vortek. Several tungsten–halogen lamp configurations were implemented in these RTP units [7]—these included linear lamps and multi-zone configuration of lamps in circular geometry. This resulted in demands on thermal management of the lamp/processing chamber as well as the need for real-time monitoring of temperature using pyrometers [8] that operated in a range of wavelengths in the infrared. Vortek proposed a single-lamp approach. These technologies continue to be practiced in the industry today.

As new materials begin to evolve, the radiative properties of semiconductors will continue to be of enormous interest to material scientists, engineers and technologists. The challenges will be to address issues such as surface/interface roughness, free carrier absorption in bulk/multi-layers as well as in two-dimensional materials and structures [9, 10]. Some efforts have been made in the literature to address surface roughness using the bidirectional reflectance distribution function (BRDF) via Monte Carlo simulations [11].

Radiative properties of semiconductors have direct applications in the design and fabrication of materials, devices, circuits and systems, particularly in the infrared, such as bolometers/microbolometers [12, 13], coatings, detectors/imagers, filters, lenses, light emitting diodes, photonic crystals, selective absorbers/emitters and solar cells/thermophotovoltaics [14]. They are critical for real-time process monitoring and control in manufacturing.

Next year, DARPA celebrates sixty years since its inception in 1958 [15]. Its efforts to bring convergence to a variety of technologies and products of interest to the Department of Defense and the civilian sector have been unique and unparalleled. At the same time, DARPA has been the major force in minimizing the time of transition from fundamental R&D to manufacturing to product availability in the market.

In putting together the various components of this overview, I thank NSF I-Corps (Michael A Ehrlich, Judith A Sheft), US Army Research Office, US Air Force Office of Scientific Research/Wright-Patterson Air Force Base, the US Department of Defense, DARPA, US Department of Energy/National Renewable Energy Laboratory, New Jersey Commission on Science and Technology (Jay Brandinger), Semiconductor Research Corporation and Sematech for the financial as well as the intellectual support.

Much of the write-up of this Foreword has taken place during flights between Newark, NJ and Dallas, TX. I thank Russ George for sharing his work on the oceans [16] and Randell Mill's work on the SunCell [17] and magnetic polymers [18] during one of these flights.

N M Ravindra (Ravi)
Newark, New Jersey
10 May 2017

References

[1] See for example; Wexler J M 1992 Factory Automation Plan Nears Completion *Computer Industry*, April 6th, 1992
Butler D H Jr 1992 SETA Support for the DARPA microelectronics technology insertion program of the Microelectronics Technology Office *Quarterly Technical Report*, For the Period 1 August 1992 to 31 October 1992, AD-A277 686, Booz Allen & Hamilton Inc.

[2] Fargher H E and Smith R A 1992 Planning for the semiconductor manufacturer of the future http://www.aaai.org/Papers/Symposia/Spring/1992/SS-92-01/SS92-01-013.pdf (accessed 10 May 2017)

[3] Harmon P 2002 The Texas Instruments MMST BPR project *Business Process Trends*

[4] Kosonocky W F, Kaplinsky M B, McCaffrey N J, Hou E S H, Manikopoulos C N, Ravindra N M, Belikov S, Li J and Patel V 1994 Multiwavelength Imaging Pyrometer *Proc. SPIE 2225, Infrared Detectors and Focal Plane Arrays III*, 26 *(July* 15, 1994)

[5] Ravindra N M, Tong F M, Schmidt W and Tello A M 1996 Spectral emissometer—a novel diagnostic tool for semiconductor manufacturing *Proc. of ISSM 1996 (Tokyo, 2–4 October 1996)*
Taylor R E and Dewitt D P 1981 Spectral emissivity at high temperatures, TPRL 250, *Final Report for AFOSR 77-3280*, AFOSR- TR. 8 1 -0 5 76

[6] http://www.afrinc.com/ (accessed 10 May 2017)

[7] Lojek B 1999 Early history of rapid thermal processing http://kutol.narod.ru/PUBL/rtp_otkr.pdf (accessed 23 May 2017)

[8] Reichel D 2015 *Temperature Measurement During Millisecond Annealing—Ripple Pyrometry for Flash Lamp Annealers, MATTWERK* (New York: Springer)

[9] Ravindra N M, Abedrabbo S, Gokce O H, Tong F M, Patel A, Velgapudi R, Williamson G D and Maszara W P 1998 Radiative properties of SIMOX *IEEE Trans. Compon. Packag. Manuf. Technol.* A **21** 441–9

[10] Muley S V and Ravindra N M 2014 Emissivity of electronic materials, coatings, and structures *JOM* **66** 616–36

[11] Zhou Y H and Zhang J M 2003 Radiative properties of semitransparent silicon wafers with rough surface *J. Heat Transfer* **125** 462–70
Shen Y J, Zhang Z M, Tsai B K and DeWitt D P 2001 Bidirectional reflectance distribution function of rough silicon wafers *Int. J. Thermophys.* **22** 1311–26

[12] Lamsal C and Ravindra N M 2014 Simulation of spectral emissivity of vanadium oxides (VOx)-based microbolometer structures *Emerging Mater. Res.* **3** 194–202

[13] Bañobre A and Ravindra N M 2016 p-silicon based microbolometer *Proc., MS&T 2016 (Salt Lake City)*

[14] Ravindra N M, Pierce D E, Guazzoni G, Babladi M and Gokce O H 2000 Radiative properties of materials of interest to thermophotovoltaics *Proc. of the 10th Workshop on Crystalline Silicon Solar Cell Materials and Processes (Copper Mountain, CO, August 2000)* 174–81

[15] http://www.darpa.mil/about-us/darpa-history-and-timeline (accessed 10 May 2017)

[16] http://russgeorge.net/ (accessed 23 May 2017)

[17] http://www.infinite-energy.com/iemagazine/issue130/WallIE130.pdf (accessed 23 May 2017)

[18] Mills R L Inorganic hydrogen and hydrogen polymer compounds and applications thereof, WO 2000007931 A2 https://www.google.com/patents/WO2000007931A2?cl=en (accessed 23 May 2017)

Acknowledgments

The authors acknowledge with thanks the financial support of DARPA, US Department of Defense, Army Research Office, US Department of Energy/ National Renewable Energy Laboratory, New Jersey Commission on Science and Technology, Semiconductor Research Corporation (and materials/wafers from its member companies – AMD, Applied Materials, IBM, Intel and Lucent) and Sematech. The authors are particularly thankful to the US Department of Energy/National Renewable Energy Laboratory for the 20+ years of continued and sustained funding. The intellectual interactions with Sufian Abderabbo (U. Jordan), Anthony Fiory (Lucent) and Bhushan Sopori (National Renewable Energy Laboratory) have been enlightening and engaging. The authors acknowledge with thanks the excellent support of Mr. Bruce Slutsky of the NJIT Library.

Author biographies

N M Ravindra

N M Ravindra (Ravi) is Professor of Physics at the New Jersey Institute of Technology (NJIT). He was the Chair of the Physics Department (2009–13) and Director, Interdisciplinary Program in Materials Science and Engineering at NJIT (2009–2016).

Ravi is the Editor-in-Chief of *Emerging Materials Research* (www.icevirtuallibrary.com/content/serial/emr). He is Series Editor of *Emerging Materials: Processing, Performance and Applications*, Momentum Press (www.momentumpress.net/). He has been a frequent Guest Editor of *JEM*, the *Journal of Electronic Materials* (www.tms.org/pubs/journals/JEM/jem.aspx); *JOM* (www.tms.org/pubs/journals/JOM/JOMhome.aspx). He serves on the editorial board of several international journals and book series that are dedicated to materials science and engineering.

Before joining NJIT in 1987, Ravi had been associated with Vanderbilt University, the Microelectronics Center of North Carolina (MCNC), North Carolina State University, International Center for Theoretical Physics (ICTP-Trieste), Politecnico di Torino, CNRS associated labs in Paris and Montpellier.

Ravi holds a PhD in Physics from Indian Institute of Technology (Roorkee, India), MS & BS in Physics from Bangalore University, India. Ravi and his research team have published over 300 papers in international journals, books and conference proceedings; his team has several pending and two issued patents; he has organized over 30 international conferences; and he has given over 75 talks in international meetings.

His research activities have been sponsored by agencies including the US Department of Defense (DOD), Defense Advanced Research Projects Agency (DARPA), SEMATECH, Semiconductor Research Corporation, US Department of Energy/National Renewable Energy Laboratory (DOE/NREL), US Department of Education, National Aeronautics & Space Administration (NASA), US Army Research Office, US Air Force of Scientific Research, New Jersey Commission on Science and Technology and the National Science Foundation.

Ravi's research interests include education, energy, health, materials, manufacturing and the Physics of Sports. He has been a frequent keynote speaker in several international conferences and has won several awards in the US and abroad.

Ravi is the co-editor of Transient Thermal Processing Techniques in Electronic Materials (TMS—The Minerals, Metals, Materials Society, 1997); he is the co-author of *Black Silicon: Processing, Properties and Applications* (Momentum Press, 2016).

Here are some of the links to Ravi's Papers etc:

https://scholar.google.com/citations?user=MNoHLhys8zkC&hl=en
https://www.researchgate.net/profile/Nuggehalli_Ravindra
https://njit.academia.edu/RavindraNuggehalli

Sita Rajyalaxmi Marthi

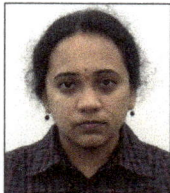

Sita Rajyalaxmi Marthi (Laxmi) received a BSc in Physics and MSc in Applied Physics from Osmania University, Hyderabad, India, in 2006 and 2009, respectively. Laxmi is currently pursuing a PhD in Materials Science and Engineering at the New Jersey Institute of Technology. Her research interests are related to the optical properties of Black silicon, related materials and device structures. Laxmi is a co-author of *Black Silicon: Processing, Properties and Applications* (Momentum Press, 2016).

Asahel Bañobre

Asahel Bañobre received his BS with Honors and MS in Applied Physics from the New Jersey Institute of Technology (NJIT) in 2003 and 2006, respectively. From 2005 to 2015, Asahel worked as a Test Engineer at Kulite Semiconductor Inc. where he contributed to the optimization and automation of device test procedures and the development of custom product prototypes. Currently, he is pursuing his PhD in Materials Science and Engineering at NJIT. His doctoral study focuses on the design, modeling, fabrication and characterization of uncooled infrared microbolometers. His research interests include semiconductors, MEMS integrated sensors and transducers such as pressure and temperature sensors, and infrared detectors.

IOP Concise Physics

Radiative Properties of Semiconductors

N M Ravindra, Sita Rajyalaxmi Marthi and Asahel Bañobre

Chapter 1

Introduction to radiative properties

1.1 Introduction

Radiative properties are fundamental physical properties that describe the interaction of electromagnetic waves, ranging from ultraviolet (UV) to deep infrared (IR) spectral regions, with matter. They can be broadly classified into optical radiative properties and thermal radiative properties. Understanding the radiative properties of materials is critical for the proper selection and interpretation of temperature measurement techniques in various industrial processes. Radiative property measurements enable us to understand the physics of solids and other states of matter. A schematic of the electromagnetic radiation spectrum is presented in figure 1.1.

Heat transfer requires a medium, either a solid or liquid. Heat transfer takes place mainly by two methods, conduction and convection. However, when heat is

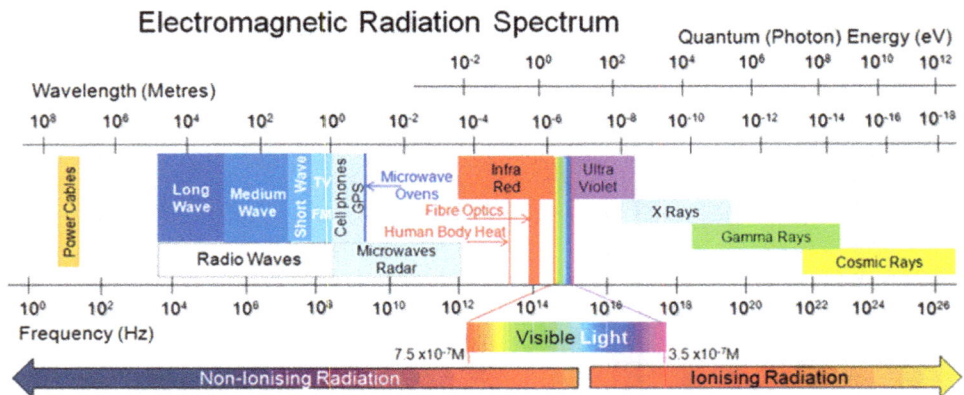

Figure 1.1. Electromagnetic radiation wave spectrum. Reproduced with permission from [1].

doi:10.1088/978-1-6817-4112-3ch1 1-1

transferred by radiation, no medium is required. Thermal radiation covers the range of wavelengths from 0.1 to 100 microns. Radiation travels through vacuum with speed of light (c) in the form of electromagnetic waves. The speed of propagation is related to its wavelength (λ) and frequency (v) by the equation:

$$c = \lambda v. \tag{1.1}$$

Frequency is independent of the medium of propagation, but velocity is dependent on the medium of propagation.

1.2 Properties

1.2.1 Emissivity

Radiation emitted by a surface is determined by introducing emissivity. Emissivity of a surface is its potential to emit energy in the form of radiation in comparison to blackbody radiation at the same given temperature. Emissivity ranges between 0 and 1.

Emissivity of a surface is given by the relation:

$$\varepsilon = \frac{q^{(e)}}{q_b^{(e)}} \tag{1.2}$$

where,

$q^{(e)}$ – energy emitted per unit area by a real body

$q_b^{(e)}$ – energy emitted per unit area by blackbody

Emissivity of black body is 1 and for real body, emissivity is <1.

Emissivity is dependent on wavelength, temperature and angle of emission of radiation. The spectral emissivity is given by:

$$\varepsilon_{\lambda,\theta}(\lambda,\, \theta,\, \phi,\, T) \equiv \frac{I_{\lambda,e}(\lambda,\, \theta,\, \phi,\, T)}{I_{\lambda,b}(\lambda,\, T)}. \tag{1.3}$$

The spectral hemispherical emissivity (directional average) is:

$$\varepsilon_\lambda(\lambda,\, T) \equiv \frac{E_\lambda(\lambda,\, T)}{E_{\lambda,b}(\lambda,\, T)} = \frac{\int_0^{2\pi} \int_0^{\pi/2} I_{\lambda,e}(\lambda,\, \theta,\, \phi,\, T) \cos\theta \, \sin\theta \, d\theta \, d\phi}{\int_0^{2\pi} \int_0^{\pi/2} I_{\lambda,b}(\lambda,\, T) \cos\theta \, \sin\theta \, d\theta \, d\phi}. \tag{1.4}$$

The total hemispherical emissivity (a directional and spectral average) is:

$$\varepsilon(T) \equiv \frac{E(T)}{E_b(T)} = \frac{\int_0^\infty \varepsilon_\lambda(\lambda,\, T) E_{\lambda,b}(\lambda,\, T) \, d\lambda}{E_b(T)}. \tag{1.5}$$

Emissivity of oxidized metal is greater than that of polished metals. Emissivity of non-conductors is relatively high. A table of emissivity of a few common materials is presented in table 1.1.

Table 1.1. Emissivity of a few materials [2].

Material	Temperature (°C)	Emissivity
Aluminium (polished)	100	0.095
Aluminium (oxidized)	200	0.11
Aluminium oxide	500–827	0.42–0.26
Bismuth, unoxidized	25	0.048
Bismuth, unoxidized	100	0.061
Graphite	0–3600	0.70–0.80
Earthenware Ceramic	20	0.90
Smooth glass	0–200	0.95
Smooth glass	250–1000	0.87–0.72
Iron, unoxidized	100	0.21
Iron, oxidized	200–600	0.64–0.78
Iron oxide	500–1200	0.85–0.89
Plaster	0–200	0.91
Zinc, highly polished	200–300	0.04–0.05
Zinc, oxidized	24	0.280
Zirconium silicate	238–500	0.920–0.800
Zirconium silicate	500–832	0.800–0.520

There are three responses to incident radiation (G_λ) on a surface:
- Reflection from surface ($G_{\lambda,\text{ref}}$)
- Transmission through surface ($G_{\lambda,\text{tr}}$)
- Absorption by surface ($G_{\lambda,\text{abs}}$)

$$G_\lambda = G_{\lambda,\text{ref}} + G_{\lambda,\text{tr}} + G_{\lambda,\text{abs}}. \qquad (1.6)$$

1.2.2 Spectral absorptivity

Thermal radiation that is incident on the surface of an opaque solid is either absorbed or reflected. The absorptivity is defined as the fraction of the incident radiation that is absorbed. For semi-transparent solids, thermal radiation impinging on the surface is absorbed, reflected or transmitted, as shown in figure 1.2.

Assuming temperature independence, spectral absorptivity is:

$$\alpha_{\lambda,\theta}(\lambda, \theta, \phi) \equiv \frac{I_{\lambda,i,\text{abs}}(\lambda, \theta, \phi)}{I_{\lambda,i}(\lambda, \theta, \phi)}. \qquad (1.7)$$

Spectral hemispherical absorptivity is:

$$\alpha_\lambda(\lambda) \equiv \frac{G_{\lambda,\text{abs}}(\lambda)}{G_\lambda(\lambda)} = \frac{\int_0^{2\pi} \int_0^{\pi/2} \alpha_{\lambda,\theta}(\lambda, \theta, \phi) I_{\lambda,i}(\lambda, \theta, \phi) \cos\theta \sin\theta \, d\theta \, d\phi}{\int_0^{2\pi} \int_0^{\pi/2} I_{\lambda,i}(\lambda, \theta, \phi) \cos\theta \sin\theta \, d\theta \, d\phi}. \qquad (1.8)$$

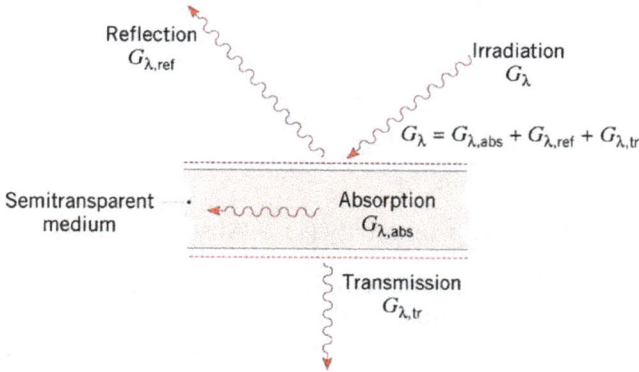

Figure 1.2. Various responses to surface irradiation. Reprinted from [3] with permission from John Wiley & Sons.

The total absorptivity is:

$$\alpha \equiv \frac{G_{abs}}{G} = \frac{\int_o^\infty \alpha_\lambda(\lambda)G_\lambda(\lambda)\,d\lambda}{\int_0^\infty G_\lambda(\lambda)\,d\lambda}. \tag{1.9}$$

1.2.3 Spectral reflectivity

Spectral directional reflectivity of a material is:

$$\rho_{\lambda,\theta}(\lambda,\,\theta,\,\phi) \equiv \frac{I_{\lambda,i,ref}(\lambda,\,\theta,\,\phi)}{I_{\lambda,i}(\lambda,\,\theta,\,\phi)}. \tag{1.10}$$

The spectral hemispherical reflectivity is:

$$\rho_\lambda \equiv \frac{G_{\lambda,ref}(\lambda)}{G_\lambda(\lambda)} = \frac{\int_0^{2\pi}\int_0^{\pi/2}\rho_{\lambda,\theta}(\lambda,\,\theta,\,\phi)I_{\lambda,E}(\lambda,\,\theta,\,\phi)\cos\theta\,\sin\theta\,d\theta\,d\phi}{I_{\lambda,i}(\lambda,\,\theta,\,\phi)}. \tag{1.11}$$

The total spectral reflectivity is:

$$\rho \equiv \frac{G_{abs}}{G} = \frac{\int_0^\infty \rho_\lambda(\lambda)G_\lambda(\lambda)\,d\lambda}{\int_0^\infty G_\lambda(\lambda)\,d\lambda}. \tag{1.12}$$

1.2.4 Spectral transmissivity

Assuming negligible temperature dependence, spectral, directional transmissivity is given by:

$$\tau_\lambda \equiv \frac{G_{\lambda,tr}(\lambda)}{G_\lambda(\lambda)}. \tag{1.13}$$

The total hemispherical transmissivity is:

$$\tau \equiv \frac{G_{tr}}{G} = \frac{\int_0^\infty G_{\lambda,tr}(\lambda)\, d\lambda}{\int_0^\infty G_\lambda(\lambda)\, d\lambda}. \tag{1.14}$$

For a semi-transparent medium,

Reflectivity (ρ) + Transmissivity (τ) + Absorptivity $(\alpha) = 1.$ (1.15)

Kirchhoff's law states that, for a material in thermodynamic equilibrium, the total hemispherical absorptivity is equal to the total hemispherical emissivity from the surface of the material, where irradiation of the surface corresponds to emission from a blackbody at the same temperature as the surface.

$$\varepsilon = \alpha. \tag{1.16}$$

1.2.5 Thermal radiation properties

1.2.5.1 Radiance
For a blackbody, radiance is defined as the radiant power dq from a pencil cone, from surface area dA in the direction $\hat{s}$, per unit wavelength interval per unit solid angle and per unit area projected onto the direction $\hat{s}$. Spectral radiance is mathematically given by:

$$L_\lambda(\lambda, \theta, \varphi) = \frac{dq}{dA \cos \theta\, d\Omega\, d\lambda} \tag{1.17}$$

where, $d\Omega$ is solid angle

$$d\lambda = \sin \theta\, d\theta\, d\varphi.$$

The total radiance is given by the integral over all wavelengths:

$$L = \int_0^\infty L_\lambda(\lambda)\, d\lambda. \tag{1.18}$$

The concept of radiance and solid angle are illustrated in figure 1.3. The radiative heat flux is the rate of energy flow per unit area across an element area dA, with units W m^{-2}.

$$q' = \frac{dq}{dA} = \int_{\varphi=0}^{2\pi} \int_{\theta=0}^{\pi/2} L \cos \theta \sin \theta\, d\theta\, d\varphi \tag{1.19}$$

where, dA is the area of the surface.

The radiative heat flux leaving the body solely by emission is also called emissive power.

An ideal blackbody absorbs all the incident radiation and does not reflect any radiation. At thermodynamic equilibrium, blackbody emits all the radiation at any

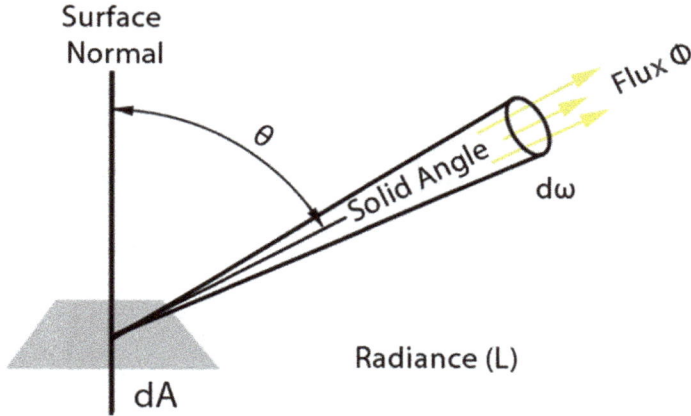

© www.scratchapixel.com

Figure 1.3. Illustration of radiance and solid angle [4].

given temperature. Blackbody is used as a reference in thermal radiation to measure the absorption of a real body.

Stefan Boltzmann law provides a relationship between the energy radiated by blackbody and its temperature. The law states that the energy radiated by black body (j) per unit area is directly proportional to the fourth power of its temperature (T).

$$j = \sigma T^4 \tag{1.20}$$

where, σ, the Stefan Boltzmann constant is equal to 5.67×10^{-8} W m^{-2} K^{-4}.

This law implies that the energy radiated by objects at higher temperatures is greater than the energy radiated by those at lower temperatures.

Planck's radiation law describes the relationship between the spectral density of the electromagnetic radiation emitted by blackbody at thermal equilibrium and temperature T. The spectral density (B_λ), in terms of wavelength, is given by the relation:

$$B(\lambda, T) = \frac{2hc^2}{\lambda^5} \frac{1}{e^{\frac{hc}{\lambda kT}} - 1}. \tag{1.21}$$

Spectral density (B_ν), in terms of frequency, is given by:

$$B(\nu, T) = \frac{2h\nu^5}{c^3} \frac{1}{e^{\frac{h\nu}{kT}} - 1}. \tag{1.22}$$

The Planck's law describes the radiation distribution that peaks at a certain wavelength or frequency. The peak shifts to shorter wavelengths or higher frequencies for higher temperatures, and the area under the curve increases with increasing temperature.

Planck's distribution curves are shown in figure 1.4. Wein's law states that the blackbody radiation curve peaks at different temperatures for a given wavelength

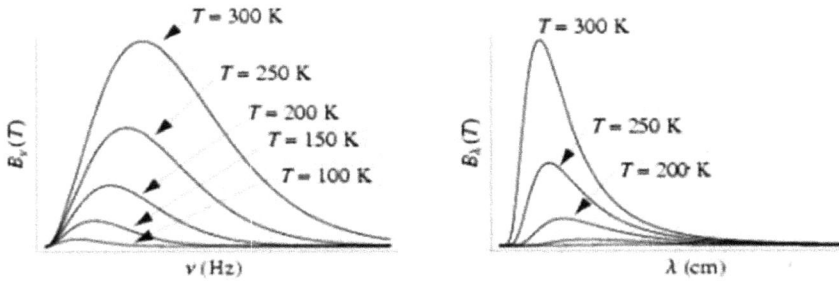

Figure 1.4. Planck's distribution [5].

and is inversely proportional to the temperature. This is as a result of Planck's law. Mathematically, Wein's law is given by the relation:

$$\lambda_{\max} = \frac{b}{T} \tag{1.23}$$

where, b is the Wein's displacement constant equal to 2.898×10^{-3} mK.

1.2.6 Radiative properties of semiconductors

Study of the radiative properties in semiconductors is very important and useful for materials and device processing. For real-time process monitoring and control, the semiconductor industry requires studies on the radiative properties of wafers. In the fabrication of advanced semiconductor devices, techniques such as rapid thermal processing (RTP) are often employed. This technique requires the understanding and modeling of thermal radiative properties of semiconductor wafers [2]. Spectral emissivity of wafer affects the amount of radiation emitted. At any given temperature, the hemispherical emissivity affects the heat loss by radiation from the wafer.

Semiconductor wafers are laminated with different types of coatings. Understanding the radiative properties of these coatings is critical for the successful implementation of pyrometers. In applications of pyrometry, the knowledge of emissivity of semiconductors, as a function of temperature and wavelength, is required. Several other factors such as surface roughness, doping, presence of coatings also influence the emissivity of the semiconductor wafer. Understanding the radiative properties under varying surface conditions and temperatures of wafers is critical in semiconductor process monitoring and control.

Semiconductors, when heated to the desired operating temperature, emit radiation. The free carriers present in the semiconductor alter the infrared absorption within the material and hence the resulting emissivity.

Spectral emissivity is a complicated function of:
    a) intrinsic emissivity of the semiconductor,
    b) extrinsic emissivity of the layers or thin films on the semiconductor,
    c) optical properties of the reflective chamber walls and lamp radiation that might transmit through the wafer and get detected by the pyrometer.

A combination of all these is defined as the effective emissivity.

References

[1] http://www.mpoweruk.com/radio.htm (accessed 27 September 2016)

[2] Table of emissivity of various surfaces, Mikron Instrument company, Inc.

[3] Incropera F P, DeWitt D P, Bergman T L and Lavine A S *Fundamentals of Heat and Mass Transfer* (New York: Wiley)

[4] http://nptel.ac.in/courses/105104100/lectureD_23/D_23_4.htm (accessed 6 November 2016)

[5] https://physics.stackexchange.com/questions/279561/radiometric-quantity-radiance-along-the-ray

Chapter 2

Optical and thermal properties

2.1 Optical properties

Optical property of a material is characterized and quantified by its interaction with the electromagnetic radiation. In semiconductors, the band structures are distinct from those of metals. These band structures result in various optical phenomena.

2.1.1 Optical conductivity

Optical conductivity implies electrical conductivity in the presence of an alternating electric field (AC field). The term 'optical' indicates the entire range of wavelengths. When photons with energy greater than or equal to the bandgap of the semiconductor are impinged on semiconductors, electron–hole pairs are generated and result in the generation of current. This process is called photoconductivity. At a certain frequency (ω), the relation between optical conductivity (σ), current density (j) and electric field (E) is given by:

$$j(\omega) = \sigma(\omega) * E(\omega). \tag{2.1}$$

The optical conductivity is most often measured in the optical frequency range via the reflectivity of polished samples under normal incidence or using variable incidence angles.

2.1.2 Heat capacity

The heat capacity of each material is different and is used to indicate the amount of heat required to raise the temperature by one degree. The exact value of the heat capacity depends on the conditions under which it is measured:
- Specific heat at constant volume (C_v)
- Specific heat at constant pressure (C_p).

doi:10.1088/978-1-6817-4112-3ch2

Heat capacity is dependent on temperature and has a limiting value of $3R$, where R is the universal gas constant ($R = 8.314$ J mol^{-1} K^{-1}).

Thermal expansion is the 'increase in dimension' of materials due to increasing vibration leading to larger inter-atomic distances, resulting from increase in average molecular kinetic energy, for materials as temperature increases.

Thermal conductivity is defined as the transfer of heat energy through a material.

$$J_Q = -\kappa \frac{dT}{dx} \qquad (2.2)$$

where J_Q is heat flux, κ is thermal conductivity and $\frac{dT}{dx}$ is temperature gradient.

Thermal conductivity involves two primary mechanisms:

- Atomic vibrations—dominant in ceramics and polymers
- Conduction by free electrons—this dominates in metals.

Initially, the thermal conductivity increases with temperature and decreases at higher temperatures. This can be attributed to the disruption of the vibration induced primary free electron conduction mechanism. Adding alloying 'impurities' also disrupts free electron conduction; alloys are therefore less conductive than pure metals. In semiconductors, the thermal conductivity is mainly controlled by scattering and defects.

At a given temperature, the thermal and electrical conductivities of metals are proportional and related to each other. The increase in temperature causes increase in thermal conductivity and decrease in electrical conductivity. This behavior is represented by the Wiedemann–Franz law.

Wiedemann–Franz law states that the ratio of the thermal conductivity (κ) to the electrical conductivity (σ) of a metal is proportional to the temperature (T). Mathematically, it is given by:

$$\frac{\kappa}{\sigma} = LT \qquad (2.3)$$

where L is the Lorenz number $= 2.44 \times 10^{-8}$ W Ω K^{-2}.

2.1.3 Drude approximation

Drude's model attributes κ in metals to the dominance of electronic contribution. A result from elementary kinetic theory is:

$$\kappa = \frac{1}{3} v_{\text{random}} l c_{\text{el}} \qquad (2.4)$$

where c_{el} is electronic specific heat per unit volume, l is mean free path and v_{random} is velocity of thermal electron.

Each electron has an energy $\frac{3}{2}k_BT$; thus, for n electrons per unit volume:

$$E_{tot} = \frac{3}{2}nk_BT, \ c_{el} = \frac{dE_{tot}}{dT} = \frac{3}{2}nk_B \tag{2.5}$$

where, mean free path l is given by:

$$l = v_{random}t, \ \text{and}$$

$$\sigma = \frac{ne^2t}{m}$$

Taking the ratio of κ to σ, and rearranging,

$$\frac{\kappa}{\sigma T} = \frac{\pi^2 k_B^2}{3e^2}. \tag{2.6}$$

Drude's model is reasonably self-consistent in identifying electron–ion collisions as the main scattering mechanism, and provides an excellent interpretation of the most universal known property of metals. In narrow bandgap semiconductors, the effective mass tensor ($m_{\alpha\beta}$, $\alpha\beta$ is rank of tensor) is a function of energy. The Drude formula is evaluated at Fermi-level for $m_{\alpha\beta}$ with n being the total carrier density.

Considering free carrier conduction mechanism alone, and all the contributions from core dielectric function (ε_{core}) to obtain the total complex dielectric function:

$$\varepsilon = \varepsilon_{core}(\omega) + \frac{4\pi i\sigma}{\omega} \tag{2.7}$$

where,
$\frac{4\pi i\sigma}{\omega}$ is the imaginary part of free carrier contribution, ω is frequency of light, i is $i = \sqrt{-1}$ and σ is conductivity,

$$\sigma(\omega) = \left(\frac{ne^2\tau}{m^*}\right)(1 - i\tau)^{-1}. \tag{2.8}$$

When the free carrier absorption is absent, $\sigma = 0$, $\varepsilon = \varepsilon_{core}$.
From Drude theory,

$$\varepsilon = (\varepsilon_1 + i\varepsilon_2) \tag{2.9}$$

$$\varepsilon = (n + ik)^2 \tag{2.10}$$

$$\varepsilon = \varepsilon_{core} + \frac{4\pi i}{\omega}\frac{ne^2\tau}{m(1 - i\omega\tau)}. \tag{2.11}$$

Further reading

[1] Optical properties, http://nptel.ac.in/courses/112108150/pdf/PPTs/MTS_17_m.pdf (accessed 11 April 2017)
[2] Thermal properties, http://nptel.ac.in/courses/112108150/pdf/PPTs/MTS_15_m.pdf (accessed 11 April 2017)
[3] Solid State Physics, Part II - Optical Properties of Solids, M Dresselhaus, http://web.mit.edu/course/6/6.732/www/6.732-pt2.pdf (accessed 11 April 2017)
[4] Hummel R E 2011 *Electronic Properties of Materials* (New York: Springer)

Radiative Properties of Semiconductors

N M Ravindra, Sita Rajyalaxmi Marthi and Asahel Bañobre

Chapter 3

Instrumentation

This chapter introduces some of the instruments that are used to measure the radiative properties of materials.

3.1 Spectral emissometer

The spectral emissometer [1, 2], shown in figure 3.1, consists of a hemiellipsoidal mirror providing two foci, one for the exciting source in the form of a diffuse radiating near-blackbody source and the other for the sample under investigation. A microprocessor controlled motorized chopper facilitates in simultaneous measurement of the sample spectral properties such as reflectivity, transmissivity and emissivity. A carefully adjusted set of five mirrors provides the optical path for the measurement of the optical properties.

The source of heating of the samples is provided by an oxy-acetylene/propane torch. However, because of safety considerations and potential sample contamination, various alternatives to heat the samples, uniformly in a controlled environment, can be implemented. The spectral emissometer consists of three GaAs lasers to facilitate in aligning the sample at the appropriate focus. A high resolution Bomem FTIR, consisting of Ge and HgCdTe detectors, interfaced with an online computer, permits data acquisition of the measured optical properties.

Further, this online computer enables the user to flip the mirrors to acquire transmission/reflection spectra via software configurations such as Spectra Calc and GRAMS. The spectral emissometer allows for simultaneous measurements of radiance, reflectance, transmittance and the temperature of the sample at the measured point. The design and operation of the spectral emissometer is based on the Helmholtz's principle of reciprocity. This principle allows one to conclude that when the incident radiation is hemispherical and the collection path is directional, it is equivalent to the incident radiation being directional and the collection path being hemispherical.

doi:10.1088/978-1-6817-4112-3ch3   3-1

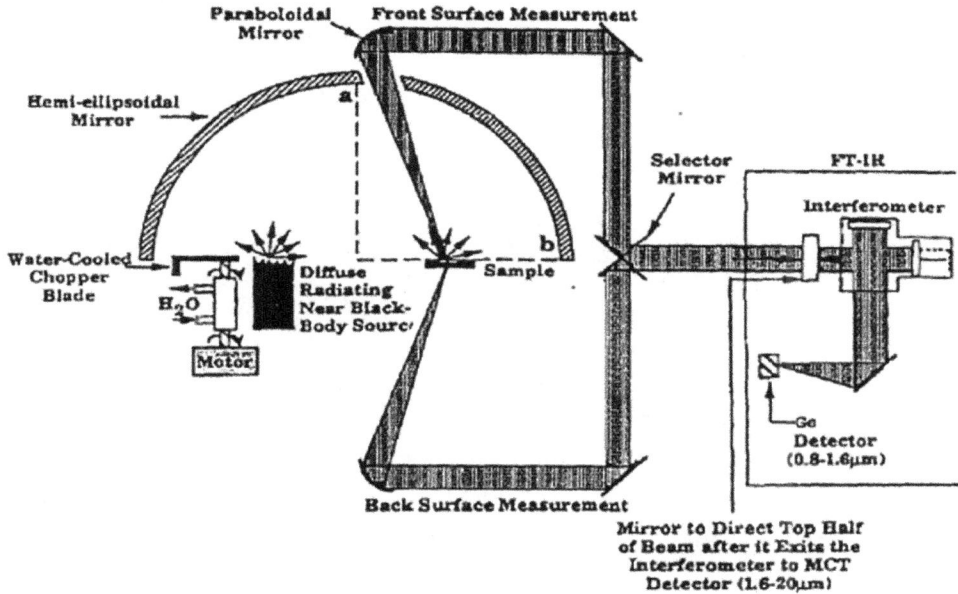

Figure 3.1. Schematic of bench top emissometer showing the components and optical paths for radiance, reflectance, and transmittance [1].

In addition to the important radiative and optical properties that the spectral emissometer acquires, it is also capable of verifying the presence of constituents or impurities in the sample as sharp features that can be observed in the wavelength range of 1 μm (10 000 cm^{-1}) to 20 μm (500 cm^{-1}). The emissometer can be used to determine the concentration of O_2 molecules.

The emissometer has the ability to detect changes in thickness of the dielectric interfaces at refractory metal base barrier, for example, W–Si–N/SiO$_2$/Si or structural changes in W–Si–N after rapid thermal annealing (RTA). It can facilitate the determination of the electron and hole masses or the optical mobilities once the absorption coefficient is known.

3.2 Czochralski crystal puller

A schematic diagram of the arrangement, used in the Czochralski (Cz) crystal puller [3], is shown in figure 3.2. The thickness of the silicon sample used is 0.5 mm. It is held in a graphite jig (consisting of two pieces of graphite, as shown in figure 3.2) in such a way that it is in intimate contact with the upper graphite block. Three 5 mm holes, one at the center and two near the periphery of the graphite block, are made. The Pt–13% Rh/Pt or chromel–alumel thermocouple junction can be inserted into any of the three holes and the temperature of a point close to the center or the periphery can be measured. The distance between the thermocouple junction and the active silicon surface is about 1 mm. It is estimated that the error in the temperature measurement due to this is a small fraction of a percent. The entire graphite/silicon

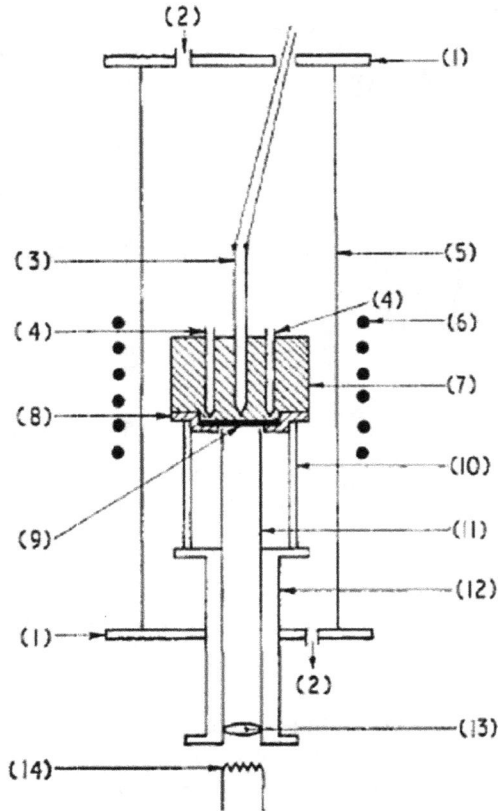

Figure 3.2. Schematic diagram for the measurement of total emissivity at high temperatures: (1) water-cooled heavy metal casing; (2) inlet and outlet for inert (nitrogen) gas; (3) Pt–13% Rh/Pt thermocouple; (4) holes for inserting thermocouples to measure the temperature of the sample near its periphery; (5) quartz envelope; (6) induction heating coil; (7) and (8) graphite jig assembly; (9) test sample; (10) quartz supporting tube; (11) translucent quartz tube; (12) molybdenum support; (13) condensing lens; (14) thermopile. Reproduced from [3] with permission of IOP Publishing.

assembly rests on a quartz tube (10) which is held in position on a molybdenum support. An inner quartz tube (11), located between the hot surface of the sample and the thermopile window, is used as a guide for radiation. It has been found that the sensitivity of the thermopile is increased by using this tube. A lens is used to focus the radiation on the thermopile to increase the sensitivity further. The measurements are made by heating the sample graphite block and silicon sample by an induction generator at 450 kHz. An inert gas (nitrogen) is passed through the quartz tube to maintain the whole graphite assembly in an inert atmosphere. The molybdenum support, the condensing lens and the thermopile are parts of the standard Czochralski crystal-pulling equipment manufactured by Kokusai Electric Company, Japan. The thermocouple output and the thermopile output are read with DC millivoltmeters. The total energy E received by the thermopile depends on the temperature of the sample, the emissivity of the surface and the geometry of the arrangement. Assuming that the total energy radiated is proportional to T^4, we may write:

$$\frac{E}{\epsilon} = AT^4 \qquad (3.1)$$

where ϵ is the total emissivity of the surface at temperature T and A is a constant for a given geometry. Since the output of the thermopile is proportional to E, equation (3.1) becomes:

$$\frac{R}{\epsilon} = BT^4 \qquad (3.2)$$

where R is the output of the thermopile in millivolts and B is a constant. By measuring the temperature T with the help of the thermocouple, the emissivity ϵ can be calculated provided the constant B is known.

3.3 Polarized radiometer

Figure 3.3 shows an experimental system for the study of silicon wafer emissivity, including a polarized radiometer [4, 5] that measures the directional polarized emissivity and a hybrid surface temperature sensor (in figure 3.4) that measures the surface temperature of a silicon wafer. The radiance emitted from an area of 2 mm in diameter, located at the center of the specimen surface, is divided into p- and s-polarized radiance by a polarizing beam-splitter cube, transmitted through optical fibers, and finally detected by respective polarized radiometers (Si sensor, IR-FBWS-SP, Chino) sensitive at the wavelength of 0.9 μm. The directional polarized emissivity, $\epsilon_{p(s)}(\theta_1)$ is the ratio of the specimen radiance to that of a blackbody at

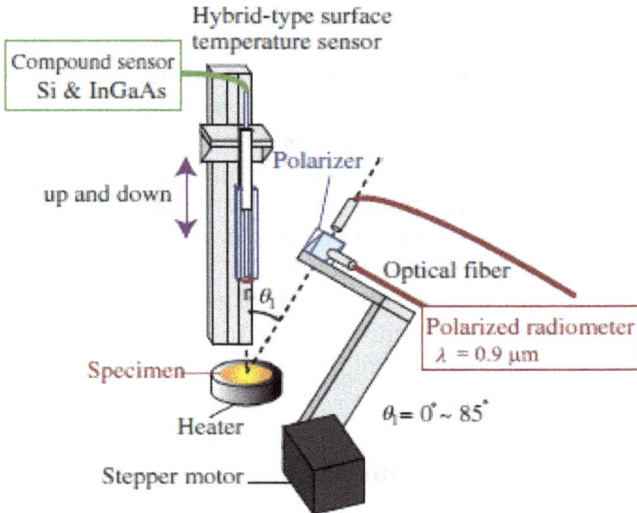

Figure 3.3. Experimental setup for measurement of polarized emissivity. Reprinted from [5] copyright 2010 with permission from Elsevier.

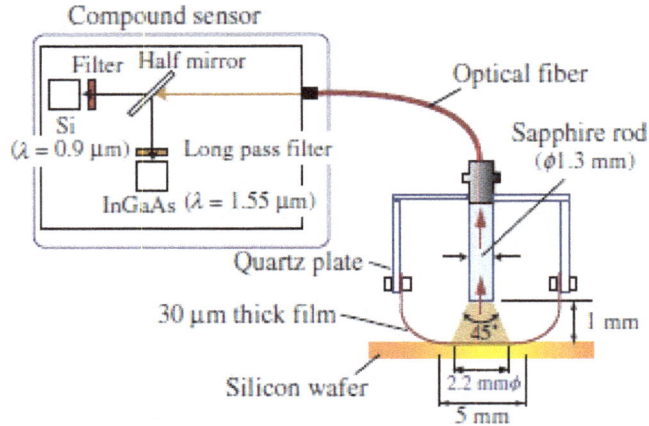

Figure 3.4. Hybrid surface temperature sensor developed for the temperature calibration of silicon wafers. Reprinted from [5] copyright 2010 with permission from Elsevier.

the same temperature; it is measured continuously at angle θ_1, between 0° and 85° in accordance with the rotation of the supporting bar. A silicon wafer is placed in the heating furnace, and maintained at a high temperature between 900 and 1000 K by the temperature controller (KP1000, Chino). The surface temperature T of the silicon wafer is monitored by a hybrid surface temperature sensor, and the p- and s-polarized radiances $E_p(T)$ and $E_s(T)$ are simultaneously measured by polarized radiometers.

Figure 3.4 shows a schematic diagram of the hybrid surface temperature sensor. This sensor is based on the assumption that the temperature of the thin film is spatially uniform at any instant during the transient process. The tip of the sensor includes a rectangular thin film of super alloy, Hastelloy (thickness: 30 μm, width: 5 mm, length: 17 mm) and a sapphire rod (diameter: 1.3 mm). Both ends of the thin metal film are supported by quartz plates, and its central portion, 5 mm in length, instantaneously contacts the surface of a silicon wafer. The thin film and the sapphire rod are closely spaced, with a gap of 1 mm. The radiant flux from the area (diameter: 2.2 mm) of the rear surface of the thin film is incident on the sapphire rod, and is transmitted through an optical fiber. The radiant flux is then detected as radiance signals by a compound sensor composed of Si and InGaAs detectors with sensitivities at 0.9 μm and 1.55 μm, respectively. When the emissivity of the rear surface of the film is known, the temperature of the film can be derived from the radiance signal. This enables accurate determination of the surface temperature of the silicon wafer. The response time of the sensor is within 1 s, and the observable temperature range is 600–1000 K.

3.4 Rotating polarizer ellipsometer

The experimental setup of a rotating polarizer ellipsometer [6] is shown in figure 3.5. A rotating polarizer ellipsometer (RPE) is attached to an ultrahigh-vacuum (UHV) chamber. Two air-spaced Glan–Taylor polarizers are employed to enhance the

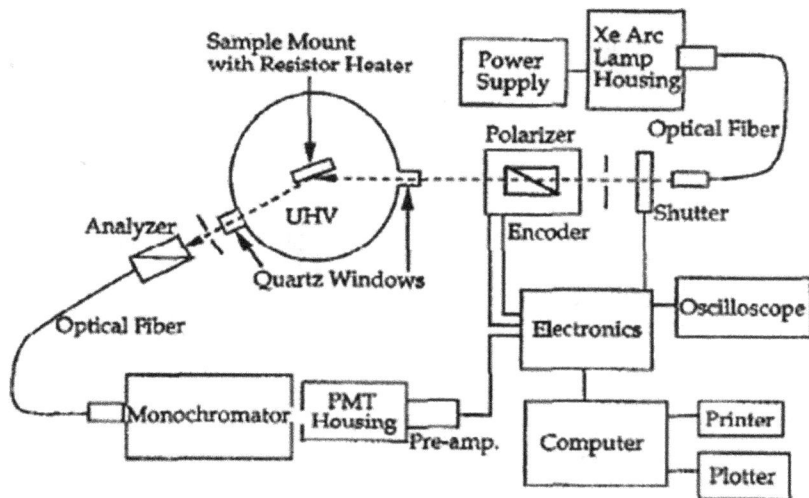

Figure 3.5. Schematic of experimental setup of rotating polarizer ellipsometer. Reproduced from [6] with the permission of AIP Publishing.

ultraviolet transmission. A monochromator is placed between the fixed analyzer and the photomultiplier tube, for improved ambient light rejection (including the blackbody radiation from the hot sample and holder). The sample is clamped on a resistor–heater plate inside the UHV chamber, which could be rotated and tilted by a rotary drive. Two low-strain, fused-quartz windows are employed in the setup. The base pressure of the UHV chamber is typically 1×10^{-9} Torr, and the spectroscopic ellipsometer measurements are carried out without arsenic overpressure.

References

[1] Abedrabbo S 1998 Emissivity measurements and modeling of silicon related materials and structures *PhD Thesis* NJIT

[2] Ravindra N M, Sopori B, Gokce O H, Cheng S X, Shenoy A, Jin L, Abedrabbo S, Chen W and Zhang Y 2001 Emissivity measurements and modeling of silicon-related materials: an overview *Int. J. Thermophys.* **22** 1593–611

[3] Jain S C, Agarwal S K, Borle W N and Tate S 1971 Total emissivity of silicon at high temperatures *J. Phys. D: Appl. Phys.* **4** 1207–9

[4] Iuchi T and Gogami A 2009 Uncertainty in the temperature of silicon wafers measured by radiation thermometry based upon a polarization technique *XIX IMEKO World Congress Fundamental and Applied Metrology (Lisbon, 6–11 September 2009)* 1487–92

[5] Iuchi T and Gogami A 2010 Simultaneous measurement of emissivity and temperature of silicon wafers using a polarization technique *Measurement* **43** 645–51

[6] Yao H, Snyder P G and Woollam J A 1991 Temperature dependence of optical properties of GaAs *J. Appl. Phys.* **70** 3261–7

Chapter 4

Silicon

During the last five decades, the semiconductor industry has been faced with a dynamic trend in increasing silicon wafer dimensions and shrinking device size. This is coupled with the demands on low power, low noise, high frequency, high reliability, thermal management and low costs of the device/circuit/product.

The industry is dominated by silicon (Si). Silicon is one the most studied and most well-understood materials. In this industry, absolute temperature measurement as well as its repeatability, reproducibility and accuracy are critical. Silicon device processing methods include the following: atomic layer deposition, chemical vapor deposition (CVD), diffusion, ion implantation, lithography, physical vapor deposition (PVD), rapid thermal processing (RTP), reactive ion etching (RIE), thermal nitridation, thermal oxidation and chemical mechanical polishing.

The intrinsic emissivity of silicon is in a non-reflective, non-radiating environment [1]. Some of the major factors that influence the intrinsic emissivity of silicon include the following:

- Surface reflectivity
- Transmittance of bulk
- Temperature
- Surface roughness.

The band gap of single crystal silicon is 1.12 eV at room temperature. Silicon becomes semitransparent at wavelengths longer than 1.1 μm [1]. In some instances, the emissivity measurement of Si meets some challenges at 4.7 μm because the radiant energy from the heater penetrates the wafer and is detected by the radiometer, causing interference. Surface morphologies of wafers help to improve the light absorbing properties of the device [2]. Deposition of several layers on the surface of the wafer influences its radiative properties.

doi:10.1088/978-1-6817-4112-3ch4

Emissivity (ϵ) is defined in terms of the true reflectivity (R) and true transmittance (T_r) of a wafer [3, 4]. The equations that are used to calculate ϵ are as follows [3, 4]:

$$\epsilon = \frac{(1 - R)(1 - T_r)}{1 - R \cdot T_r} \qquad (4.1)$$

where

$$R = \frac{(n - 1)^2 + k^2}{(n + 1)^2 + k^2} \qquad (4.2)$$

and

$$T_r = \exp(-\alpha t) = \exp(-4\pi k t / \lambda). \qquad (4.3)$$

Taking multiple reflections into consideration, the emissivity can be equated to absorptivity:

$$\epsilon = 1 - (R + T_r) \qquad (4.4)$$

Silicon has significantly higher emissivity at wavelengths beyond the absorption edge. As the temperature increases, the emissivity of Si increases. Figure 4.1 shows the spectral emissivity of p-doped silicon wafer at various temperatures. This result from Sato [5] has been the most cited work on the emissivity of silicon in the literature.

For Si, high emissivity in the visible region can be attributed to band to band transitions [5]. At low temperatures, Si is transparent for wavelengths greater than 1.2 μm, low emissivity is observed in the near-IR. Below 2 μm, the emissivity of

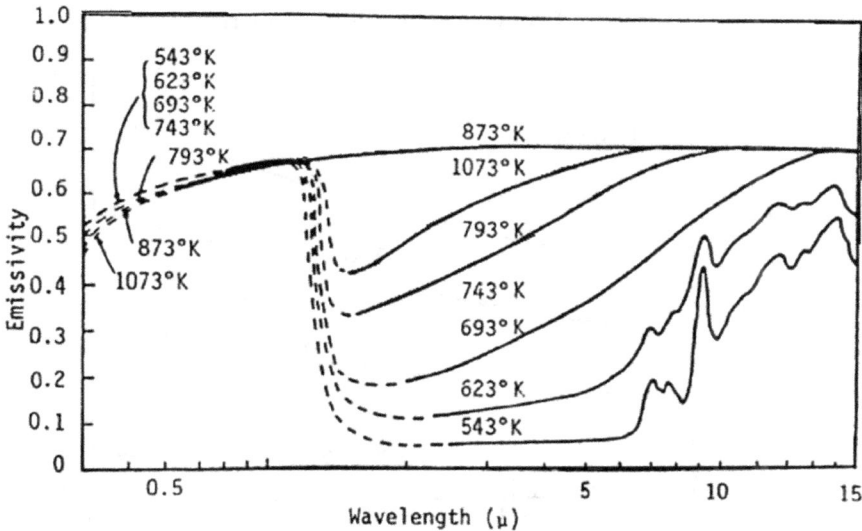

Figure 4.1. Spectral emissivity of a single-crystal n-type double-side polish silicon disc, 1.77 mm thick, doping = 2.94×10^{14} cm^{-3}, as function of wavelength and temperature. Reproduced from [5] with permission.

silicon is highly sensitive to temperature. Above 2 μm, emissivity increases with temperature irrespective of wavelength. Non-linearities in the variation of emissivity may be attributed to the presence of impurity states within the bandgap of the material and electron–phonon interaction. For wavelengths beyond 6 μm, the increase in emissivity is due to lattice vibrations. For temperatures at or above 570 K, Si becomes intrinsic leading to contribution to emissivity by free carriers. For temperatures greater than 870 K, emissivity is constant in the entire IR region. The summary of low temperature emissivity data of Si is presented in table 4.1.

Emissivity is a volume effect. As the wafer thickness increases, emissivity also increases. The emissivity of silicon increases with increase in temperature and thickness. The influence of thickness on emissivity is shown in figures 4.2–4.4.

Table 4.1. Low temperature emissivity data [6].

			Si (Sato)	
NJIT p-Si polished one side	n-Si polished both sides	n-Si polished one side	pure	n-Si polished both sides
300 K	303 K	461 K	543 K	543 K
1.4–10 μm	1.4–10 μm	1.4–10 μm	1.5–6.5 μm	2–15 μm
$\varepsilon = 0.6$–0.75	$\varepsilon \approx 0$	$\varepsilon = 0.5$–0.8	$\varepsilon \approx 0.05$	$\varepsilon = 0.7$–0.8
$\rho = 1$–$2 \times 10^{-2}\ \Omega$ cm	$\rho = 1$–$3\ \Omega$ cm	$\rho = 5$–$20 \times 10^{-3}\ \Omega$ cm	$\rho = 15\ \Omega$ cm	$\rho = 7 \times 10^{-3}\ \Omega$ cm
$t = 0.61$–0.64 mm	$t = 0.27$–0.29 mm	$t = 0.41$–0.46 mm	$t = 1.77$ mm	$t = 3.71$ mm

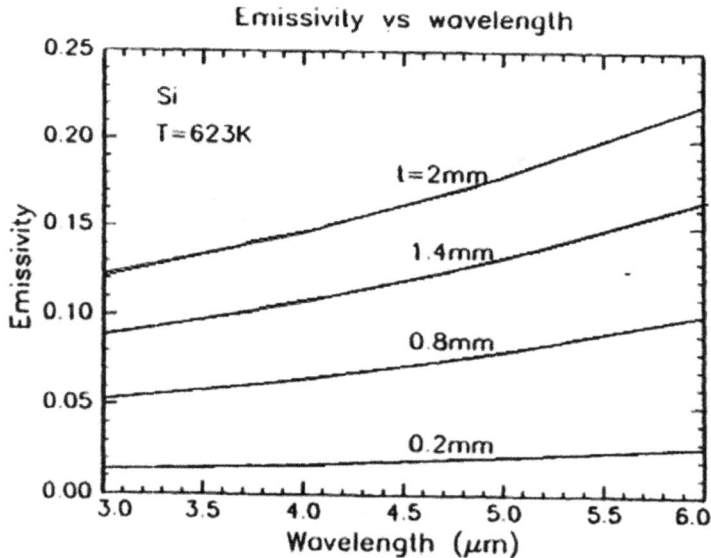

Figure 4.2. Emissivity of lightly doped silicon as a function of wavelength for various thicknesses at 623 K [6, 7].

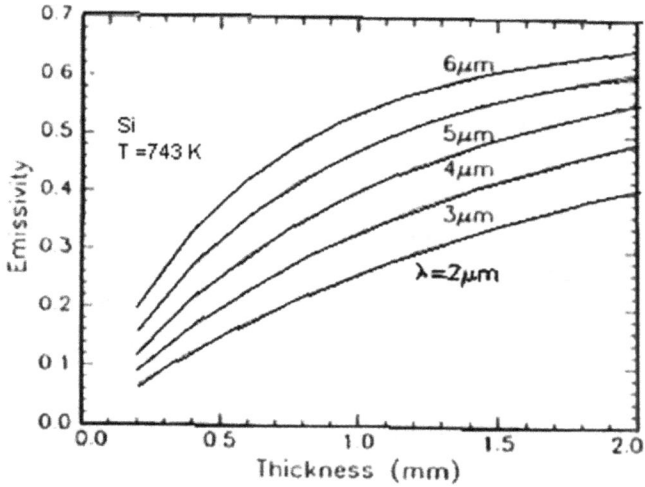

Figure 4.3. Emissivity of lightly doped silicon as a function of thickness for various wavelengths at 723 K [6, 7].

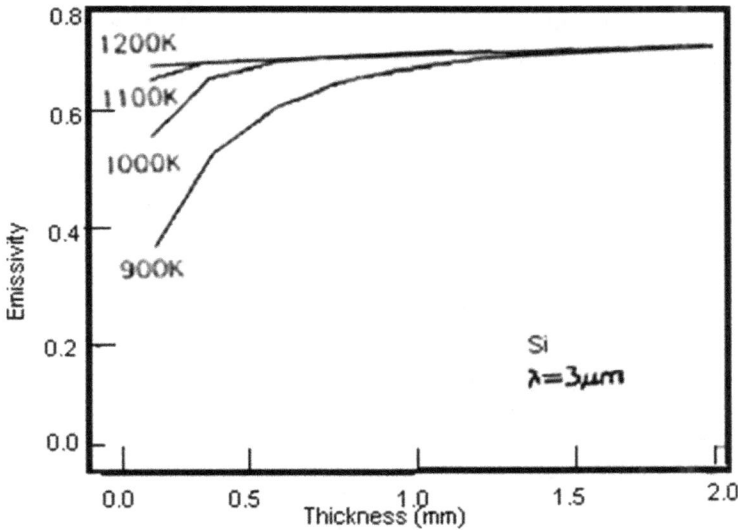

Figure 4.4. Emissivity of lightly doped silicon as a function of thickness for different temperatures at 3 μm wavelength [6, 7].

The total emissivity of lightly doped Si is strongly dependent on temperature. Increase in the total emissivity, as a function of temperature, is a result of increasing free carrier concentration [2]. The normalized spectral emissivity of lightly doped silicon is shown in figure 4.5. Free carrier concentration in silicon is a function of doping at low temperatures. It is very low in lightly doped Si, implying poor

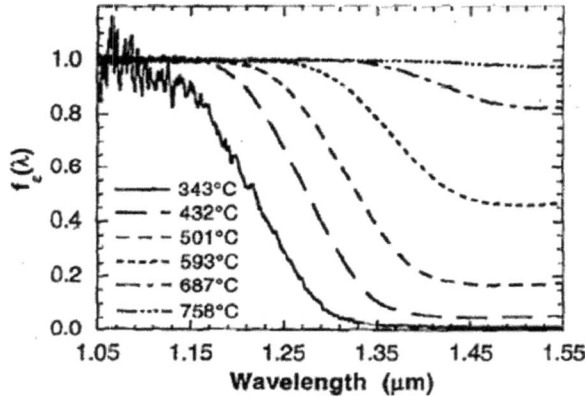

Figure 4.5. The normalized spectral emissivity $f_e(\lambda)$ of lightly doped silicon for a range of temperatures. Reprinted from [8] with the permission of AIP Publishing.

free-carrier absorption in lightly doped Si. The low total emissivity is a result of absorption associated with weak lattice vibrations [8]. Free carrier absorption, accompanied by enhancement in absorption, increases with increase in temperature, resulting in an augmentation of total emissivity. In heavily doped Si, carrier concentration is high even at room temperature. The additional thermal generation of carriers produces an insignificant change in the total emissivity.

4.1 Influence of coatings on emissivity

Coatings can have tremendous impact on the emissivity of silicon wafers [6, 7, 9]. It is critical to understand the behavior of emissivity as a function of wavelength and coating. The semiconductor industry commonly uses SiO_2, Si_3N_4 coatings on Si wafers for various applications such as gate dielectric, passivation layer etc. The emissivity of the wafer will be effectively changed upon adding a new layer of dielectric or metal or another semiconductor. For example, during the growth of a thermal oxide of 0.4 μm thickness, the reflectance of a silicon wafer, in a chamber with highly reflective walls for temperatures >600 °C, is low. For $T > 600$ °C, the emissivity increases from 0.71 to 0.87.

Dielectric materials such as SiO_2 have low losses due to their low extinction coefficients [6]. The absorption coefficient of SiO_2 is weakly dependent on temperature. The interference effect will be the dominant influencing factor on the apparent reflectance of the wafer. Thus, if the interference of the two reflected waves, one from the top and the other from the bottom planes of the oxide, is constructive, the measured reflectance will be maximum resulting in minimum emissivity, and the opposite is true. The emissivity of n-doped silicon with SiO_2 coating is shown in figure 4.6.

The spectral emissivity of the tantalum (Ta) deposited samples, at room temperature, is much lower than that of Si, in the near IR wavelength range [10]. The emissivity is higher than expected for Ta or tantalum silicide ($TaSi_2$). A thin film

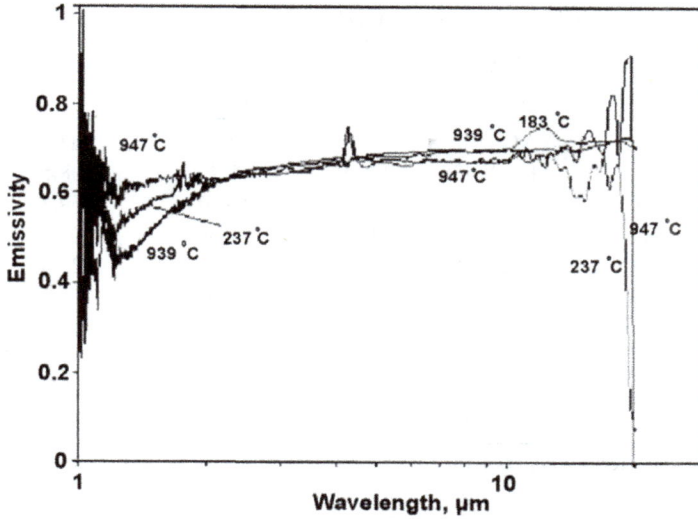

Figure 4.6. Experimental results of emissivity as function of wavelength for single-side polished n-Si of thickness 400–457 μm and doping concentration of 1.22×10^{19} to 1.31×10^{18} cm^{-3}. Reproduced from [9] with permission of Springer.

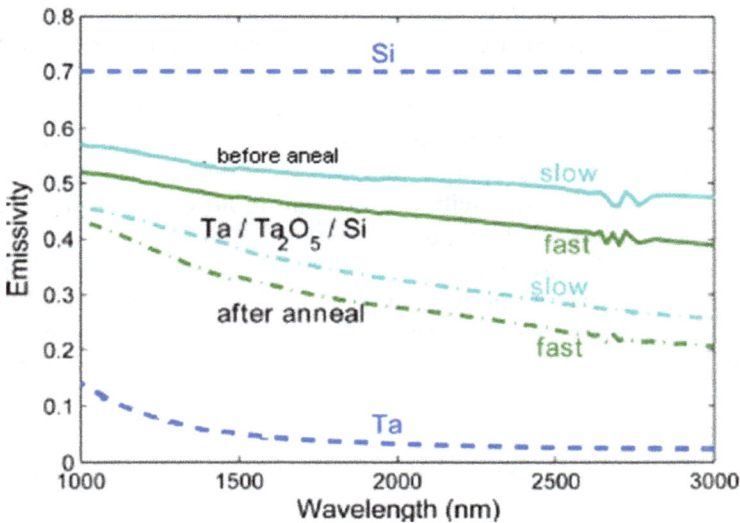

Figure 4.7. Emissivity spectra before annealing (solid lines) and after annealing at 900 °C (dashed-dotted lines), compared to emissivity of Ta and Si taken from literature. Reprinted with permission from [10] Copyright 2013, American Vacuum Society.

of Ta with thickness of ~230 nm, deposited on Si wafer, is considered in this example. Ta film causes a substantial decrease in emissivity. Emissivity is 0.28 and 0.31 for Ta layers deposited at the rate of 2 Å s^{-1} and 1 Å s^{-1}, respectively after annealing at 900 °C. Emissivity spectra of Ta coated Si wafer are presented in figure 4.7.

Table 4.2. Results of *in situ* emissivity measurements [11].

Wafer	Type	Back oxide	RTA system	Nom. temp.	Sheet result.	True temp.	Emiss. (%.)
1	#0	...	AG 610	1100	45.24	1137.6	79.0
2	#0	...	ROA 400	1100	48.15	1108.5	68.8
4	#0	400 nm	AG 610	1100	53.07	1059.3	95.0[a]
5	#0	260 nm	ROA 400	1172.5	48.44	1105.6	95.0[a]
7	#1	...	AG 610	1100	46.06	1129.4	80.4
8	#1	...	ROA 400	1100	47.98	1110.2	68.3
13	#2	...	AG 610	1100	47.44	1115.6	83.0
14	#2	...	ROA 400	1100	48.34	1106.6	69.4
21	#3	...	AG 610	1100	50.00	1090.0	88.2
20	#3	...	ROA 400	1100	50.30	1087.0	75.8

[a] The emissivity of the reference wafers was defined *a priori* as 95%.

The emissivity of an uncovered, polished Si wafer is strongly dependent on the presence of grown or deposited layers on the backside of the wafer and the backside roughness of the wafer. At low temperatures (~600 °C–700 °C), bulk doping, implanted layers, wafer thickness, and films on frontside also influence the backside emissivity because the wafer is partially transparent [8]. The emissivity is less sensitive to the exact measurement of wafer temperature. Results of *in situ* measurements of emissivity in RTP are shown in table 4.2.

A non-contact temperature measurement device, such as an optical pyrometer, operating in the wavelength range of 0.85–1.0 μm is considered here. At these wavelengths, the Si wafer is opaque and the front side emissivity is 69%. At a wavelength of 3.4 μm, the emissivity saturates at about 620 °C at a value of 68%. Emissivity is independent of the type of doping.

The wafer is partially transparent at lower temperatures. As seen in figure 4.8, at low temperature, with increasing roughness, the emissivity increases. The emissivity is increased by light trapping of scattered radiation inside partially transparent wafer [12]. The influence of roughness on emissivity may be explained by the following parameters:

- Surface reflectivity
- Correction factor for decrease of internal transmission due to light trapping.

The emissivity dependence on backside roughness is also illustrated in figure 4.9. At elevated temperatures, a small change in the refractive index is caused by changes in intrinsic and free carrier absorption [11–13]. The influence of extinction coefficient on emissivity is small even at high impurity concentrations of the order of 8×10^{18} cm^{-3}.

The emissivity of the rough side is higher than the emissivity of the polished side. The difference is evident until the temperature reaches 700 °C and the wafer is opaque to sub-band gap radiation.

Values of normal spectral emissivity are presented in table 4.3.

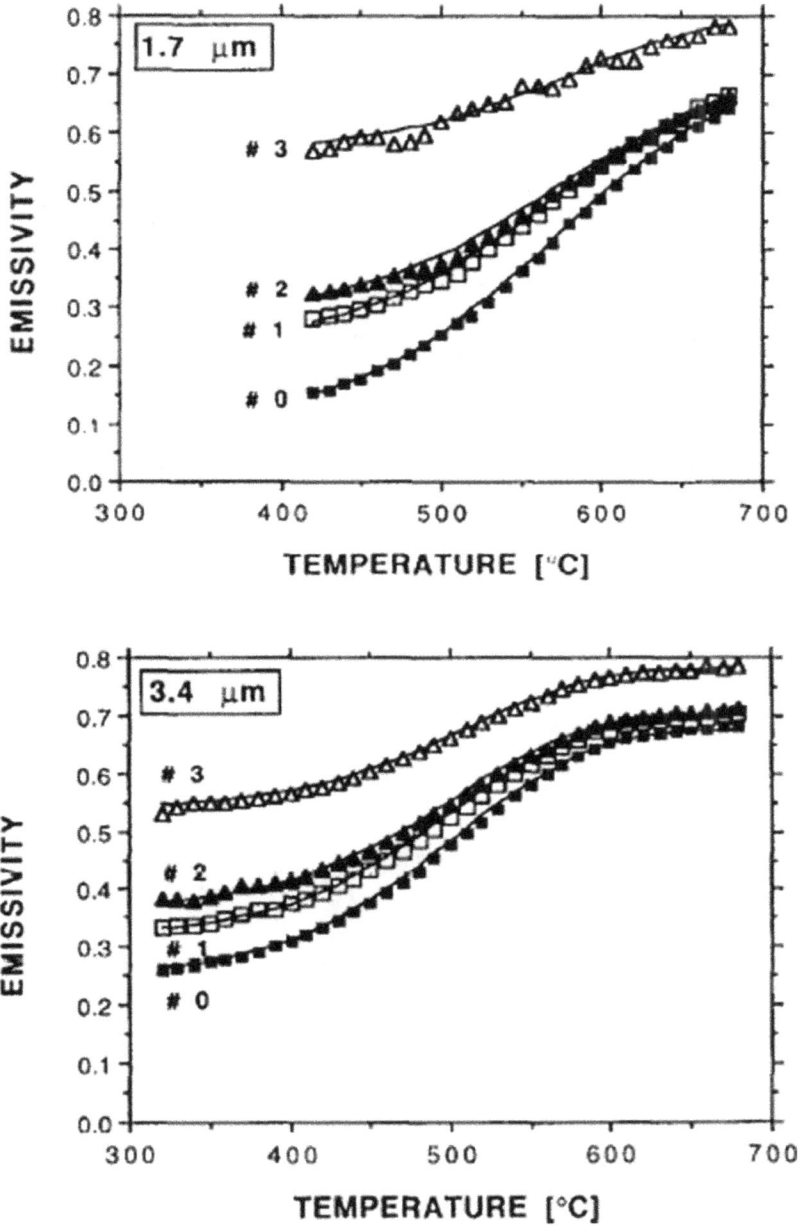

Figure 4.8. Backside emissivity of four wafers with increasing backside roughness at wavelengths of (a) 1.7 μm and (b) 3.4 μm. Reprinted from [11] with the permission of AIP Publishing.

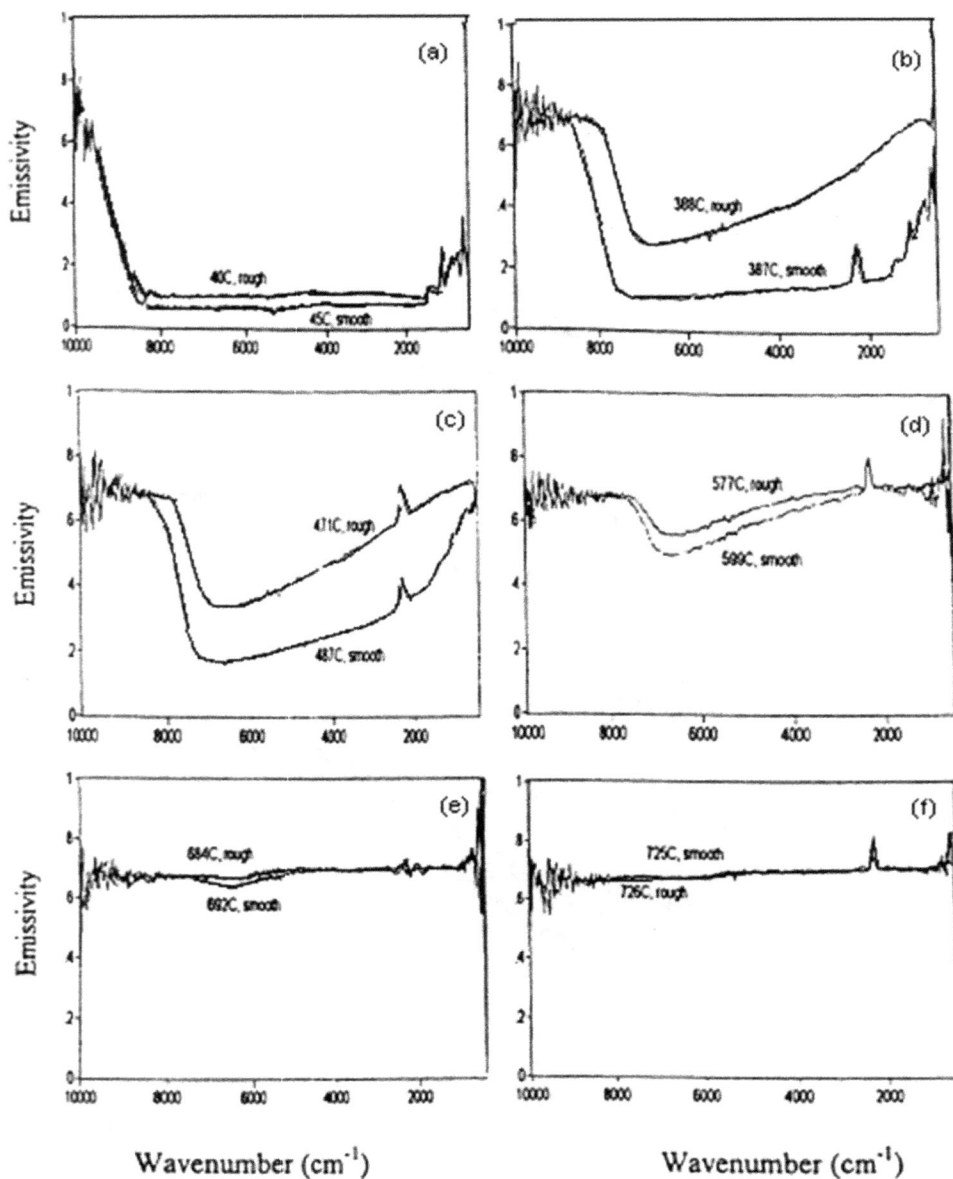

Figure 4.9. Emissivity as function of wavenumber for specific temperatures: (a) 40 °C, rough; 45 °C, smooth (b) 388 °C, rough; 387 °C, smooth (c) 471 °C, rough; 487 °C, smooth (d) 577 °C, rough; 599 °C smooth (e) 684 °C, rough; 692 °C, smooth (f) 726 °C, rough; 725 °C smooth. Reproduced with permission from [12, 13].

Table 4.3. Values of normal spectral emissivity and relaxation time at 14 microns and 1074 K [14].

Electron carrier concentration, N_n, electrons/cm^3	Normal spectral emissivity, ϵ_λ	Relaxation time, τ, s
8.5×10^{19}	0.46	0.39×10^{-14}
3.7	0.62	0.47
2.2	0.69	0.55

Hole carrier concentration, N_p, holes/cm^3	Normal spectral emissivity, ϵ_λ	Relaxation time, τ, s
14×10^{19}	0.49	0.29×10^{-14}
6.2	0.62	0.39

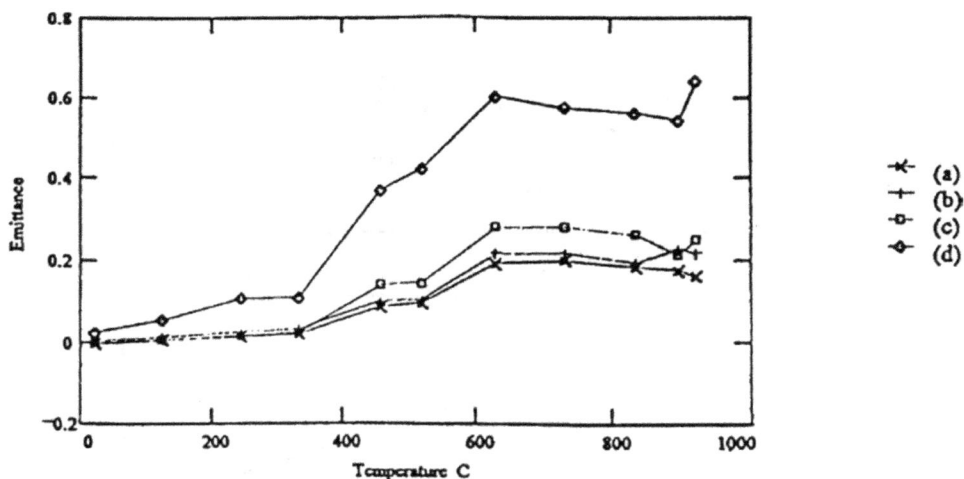

Figure 4.10. Emissivity as function of temperature for SIMOX, for λ = (a) 2.5, (b) 2.7, (c) 3.3, and (d) 4.5 μm [15].

At longer wavelengths and at high temperatures, the emissivity decreases as the doping concentration increases.

One of the methods to fabricate silicon on insulator (SOI) wafers is to implant oxygen into Si, i.e. separation by implantation of oxygen (SIMOX) [15]. The emittance of SIMOX increases at higher temperatures. At 807 °C, in the wavelength range of 1.5–5 μm, free carrier absorption mechanism is dominant in Si. SIMOX differs from Si as the buried oxide reflects the IR photons in this range of wavelengths, decreasing the emissivity. With increase in temperature, the emissivity of SIMOX is almost comparable to that of Si. The emissivity of SIMOX, as a function of temperature, for four selected wavelengths is shown in figure 4.10. As can be seen in this figure, within the range of wavelengths considered, the emissivity at 4.5 μm is the maximum and approaches 0.7 at 900 °C.

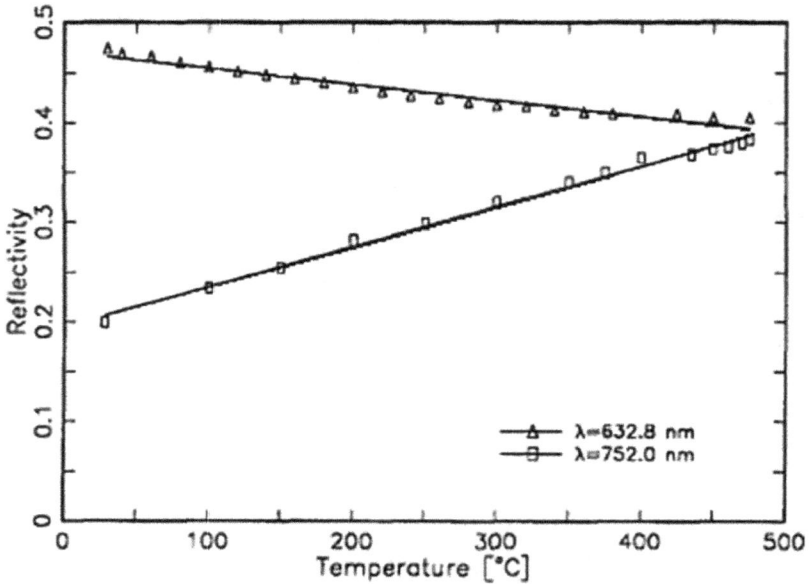

Figure 4.11. Reflectivity as a function of temperature for a-Si sample of 0.2 μm thickness. Reproduced from [16] with permission of the Optical Society of America.

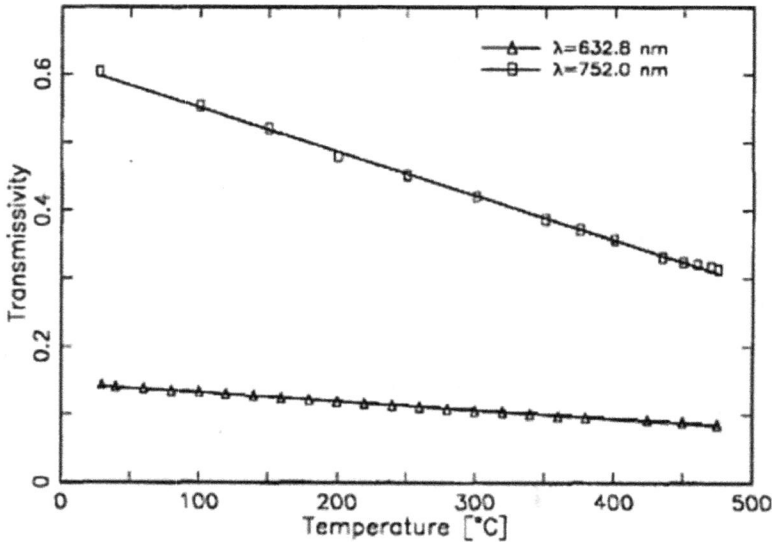

Figure 4.12. Transmissivity as a function of temperature for a-Si sample of 0.2 μm thickness. Reproduced from [16] with permission of the Optical Society of America.

The influence of temperature on the optical properties of amorphous silicon (a-Si) is a topic that has not been studied in great detail since it is believed that the optical constants of amorphous silicon are independent of temperature [16]. However, some studies show that there is a larger influence of temperature on the optical constants of a-Si, especially, in the stronger absorption region. The reflectivity of a-Si at wavelengths of 632.8 and 752.0 nm is presented in figure 4.11.

Figure 4.12 shows the dependence of transmissivity on temperature for a-Si. It can be observed in these figures that both R and T change linearly with temperature. While the reflectivity increases at a wavelength of 752 nm, it decreases for the wavelength of 632.8 nm. The behavior of transmissivity is completely opposite. The transmissivity decreases with increase in temperature for both the wavelengths.

References

[1] Timans P J 1996 The thermal radiative properties of semiconductors *Advances in Rapid Thermal and Integrated Processing* ed F Roozeboom (Berlin: Springer) pp 35–101
[2] Sugawara H, Ohkubo T, Fukushima T and Iuchi T 2003 Emissivity measurement of silicon semiconductor wafer near room temperature *SICE Annual Conference (Fukui, 4–6 August)* 2201–4
[3] Sopori B, Chen W, Madjpour J and Ravindra N M 1999 Calculation of emissivity of silicon wafers *J. Electron. Mater.* **28** 1385–9
[4] Vandenabeele P and Maex K 1990 Emissivity of silicon wafers during rapid thermal processing *SPIE* **1393** 316–36
[5] Sato T 1967 Spectral emissivity of silicon *Japan. J. Appl. Phys.* **6** 339–47
[6] Ravindra N M, Tong F M, Amin S, Shah J, Kosonocky W F, McCaffrey N J and Manikopoulos C N 1994 Development of emissivity models and induced transmission filters for multi-wavelength imaging pyrometry (M-WIP) *SPIE* **2245** 304–18
[7] Ravindra N M, Chen W, Tong F M and Nanda A 1996 Emissivity measurements and modeling—an overview *Transient Thermal properties in Electronic Materials*, The Minerals, Metals & Materials Society pp 159–64
[8] Timans P J 1993 Emissivity of silicon at elevated temperatures *J. Appl. Phys.* **74** 6353–64
[9] Ravindra N M, Sopori B, Gokce O H, Cheng S X, Shenoy A, Jin L, Abedrabbo S, Chen W and Zhang Y 2001 Emissivity measurements and modeling of silicon-related materials: an overview *Int. J. Thermophys* **22** 1593–611
[10] Rinnerbauer V, Senkevich J J, Joannopoulos J D, Soljačić M, Celanovic I, Harl R R and Rogers B R 2013 Low emissivity high-temperature tantalum thin film coatings for silicon devices *J. Vac. Sci. Technol.* A **31** 1–5
[11] Vandenabeele P and Maex K 1992 Influence of temperature and backside roughness on the emissivity of Si wafers during rapid thermal processing *J. Appl. Phys.* **72** 5867–75
[12] Abedrabbo S 1998 Emissivity measurements and modeling of silicon related materials and structures *PhD Thesis* NJIT
[13] Abedrabbo S, Hensel J C, Gokce O H, Tong F M, Sopori B, Fiory A T and Ravindra N M 1998 Issues in emissivity of silicon *Mat. Res. Soc. Symp. Proc.* **525** 95–102
[14] Liebert C H and Thomas R D 1967 Spectral emissivity of highly doped silicon *Thermophysics of Spacecraft and Planetary Bodies* (AIAA series Progress in Astronautics and Aeronautics vol 20) (Academic Press) pp 17–40

[15] Ravindra N M, Abedrabbo S, Gokce O H, Tong F, Patel A, Velagapudi R, Williamson G D and Maszara W P 1998 Radiative properties of SIMOX IEEE *Trans. on Components, Packaging, and Manufacturing Technology—Part* A **21** 441–9
[16] Yavas O, Do N, Tam A C, Leung P T, Leung W P, Park H K, Grigoropoulos C P, Boneberg J and Leiderer P 1993 Temperature dependence of optical properties for amorphous silicon at wavelengths of 632.8 and 752 nm *Opt. Lett.* **18** 540–2

Chapter 5

Germanium

Germanium (Ge) is a group four element. Germanium is a lustrous, hard, gray-white semiconductor element with a crystalline and brittle structure. Like silicon and water, Ge expands when it freezes. Germanium finds most of its use in the semiconductor industry. Lightly doped Ge is used to manufacture electronic devices such as transistors. Both Ge and germanium dioxide (GeO_2) are transparent to infrared radiation and hence may be used as infrared detectors and infrared spectroscopes. The high index of refraction and dispersion properties of its oxide have made germanium useful as a component of wide-angle camera lenses and microscope objectives. Ge has a direct band gap of about 0.66 eV.

The relationship for bandgap dependence on temperature in Ge is given by [1]:

$$E_g = 0.742 - 4.8 \cdot 10^{-4} \cdot T^2/(T + 235)(\text{eV}) \tag{5.1}$$

where, T is the temperature in degrees K.

The temperature dependence of thermal conductivity (κ) of Ge, in the low temperature range, is shown in figure 5.1. For pure specimens, κ is proportional to $T^{-1.3}$ between 200 and 40 K. As the temperature is decreased, the thermal conductivity reaches a maximum between 12 and 30 K.

With a further decrease in temperature, the conductivity decreases rapidly. The optical absorption by free electrons in germanium is an indirect process, since phonons must be present to conserve momentum in the interaction. Figure 5.2 shows the thermal conductivity of Ge as a function of temperature in the low to high temperature range.

Absorption by free holes in Ge is primarily a direct process involving transitions between the three branches of the valence band [4]. Figure 5.3 shows the temperature dependence of emission in p-type Ge. The key features are: (1) the rapid increase in

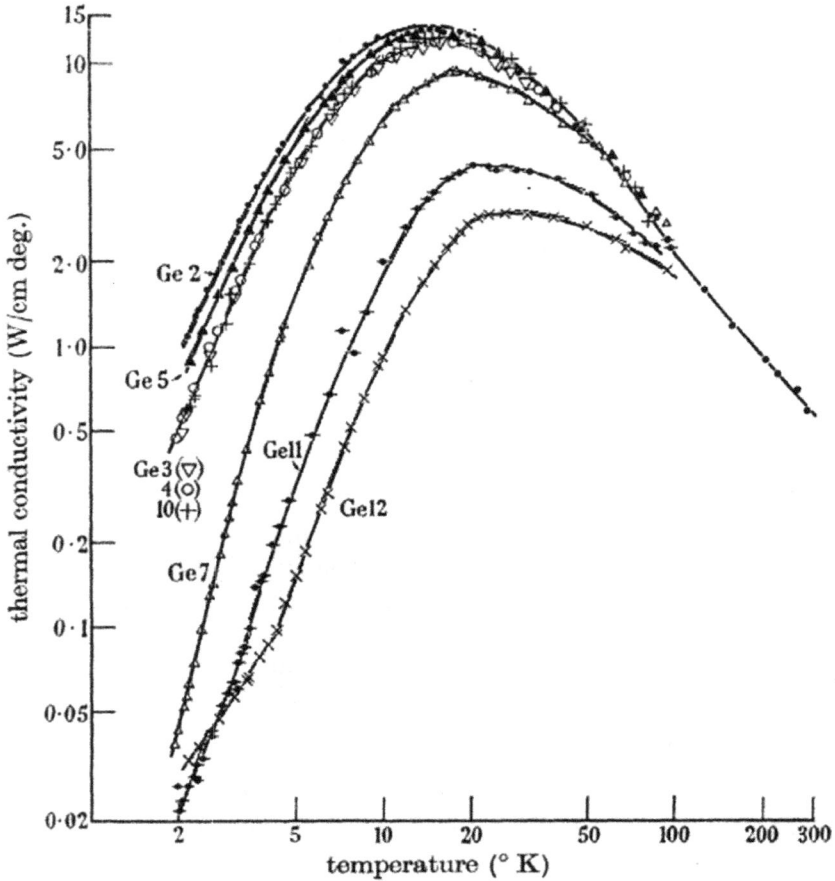

Figure 5.1. Thermal conductivity versus temperature of Ge single crystals plotted on logarithmic scale. Reproduced from [2] by permission of the Royal Society.

the emittance at higher sample temperatures, and (2) extrapolation to zero emission at room temperature.

In figure 5.4, the exitance from p-type Ge is shown as a function of the diode current for several sample temperatures. The results for n- and p-type material are very similar. The exitance increases approximately linearly with current, and is zero for zero current at all temperatures. After the initial injection of excess carriers, the exitance increases during the current pulse and then decreases approximately exponentially as the carriers recombine. The absorption and emission spectra are in accord with Kirchhoff's law. In figure 5.3, the free-carrier emission is seen to exhibit a rapid increase in exitance for temperatures above 295 K. The blackbody function, the carrier mobilities, and the absorption cross sections all contribute to the increase of the exitance at higher sample temperatures. The primary source of the

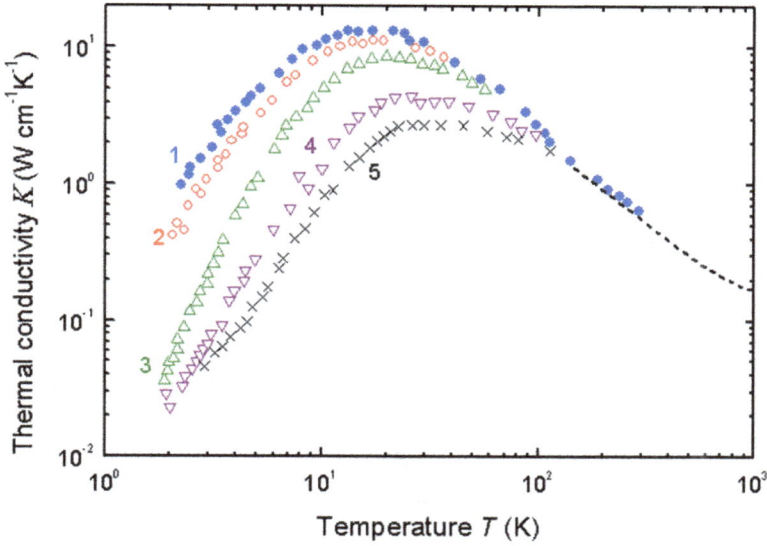

Figure 5.2. Thermal conductivity of Ge as a function of temperature plotted on logarithmic scale [3].

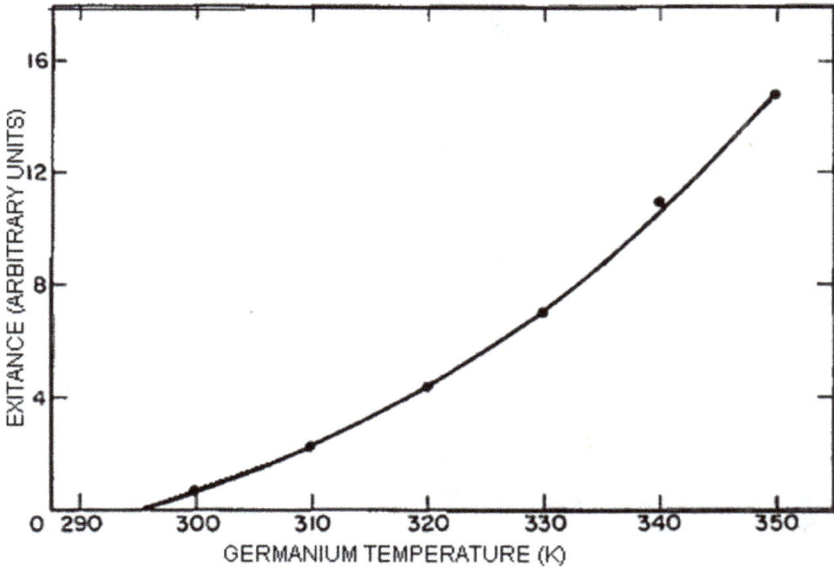

Figure 5.3. Temperature dependence of the emission in p-type Ge at a wavelength of 5.1 μm and a diode current of 600 mA. Reproduced from [4] with permission of the Optical Society of America.

rapid increase is the exponential temperature dependence of the blackbody exitance. The second most important reason for the increase is the reduced mobilities of the injected carriers, which increase the carrier concentration per unit current, and, hence, the absorption coefficient.

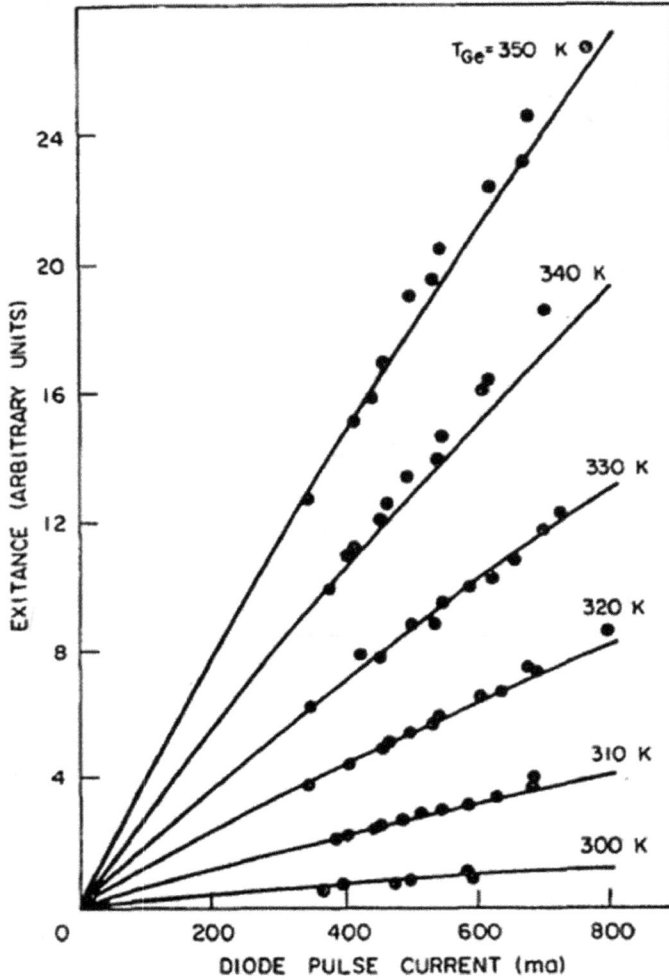

Figure 5.4. Current dependence of the emission in p-type Ge at a wavelength of 5.1 μm [4].

Emissivity of Ge at various temperatures is presented in figure 5.5. The infrared emissivity of Ge shifts at the edges at longer wavelengths by -3×10^{-4} eV deg^{-1}. The emissivity increases with increase in temperature. The long wavelength emissivity is of the order of 5% at lower temperatures, with a significant increase above 200 °C.

Figure 5.6 exhibits the specific heat of Ge as a function of temperature. The specific heat increases rapidly until about 200 °C. The specific heat is in accord with the Debye theory until the melting temperature of germanium.

Transmission of germanium, as a function of wavelength, is observed to increase at the absorption edge. The absorption edge shifts to lower energies as the temperature increases. The near and mid-wave infrared transmission is shown in figures 5.7 and 5.8, respectively.

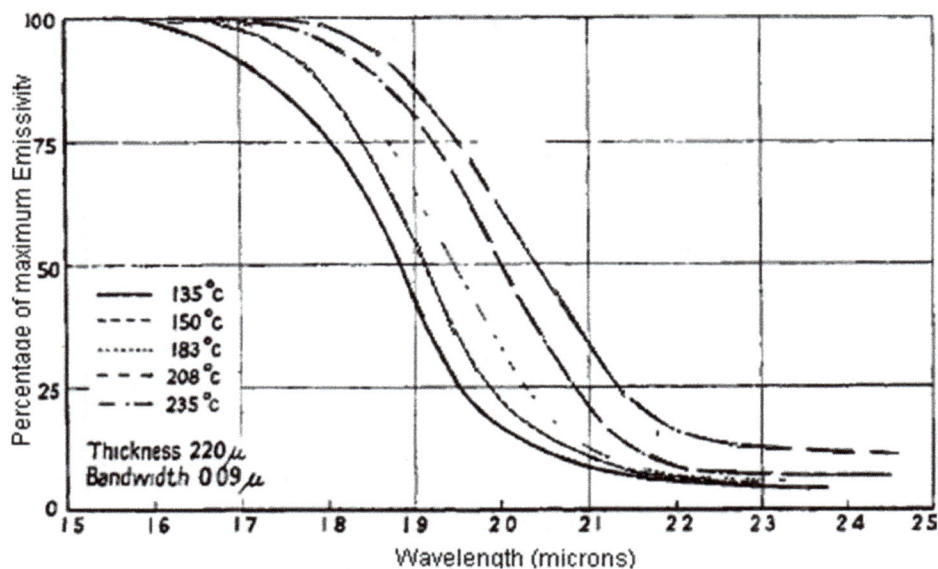

Figure 5.5. Emissivity of germanium at various temperatures. Reproduced with permission from [5].

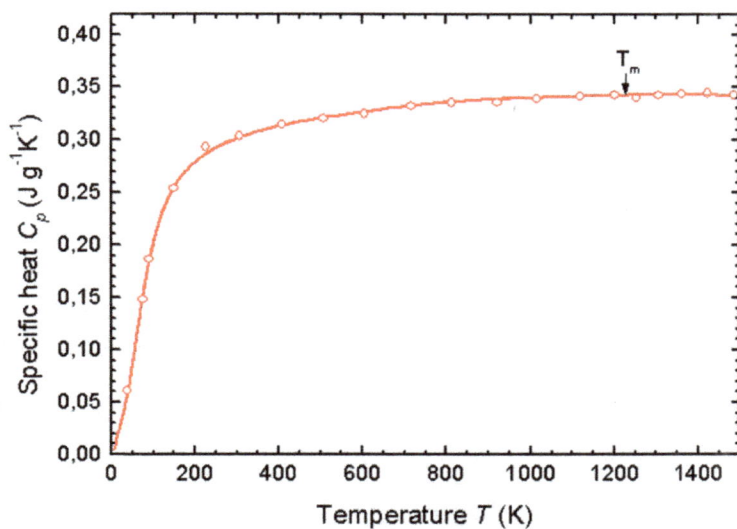

Figure 5.6. Specific heat of germanium as a function of temperature [3].

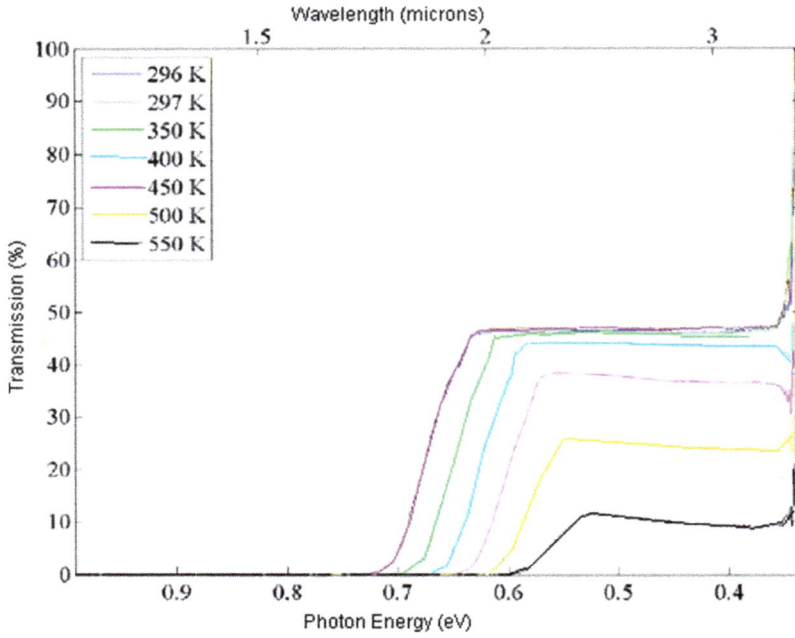

Figure 5.7. Near infrared transmission of Ge versus photon energy at various temperatures [6].

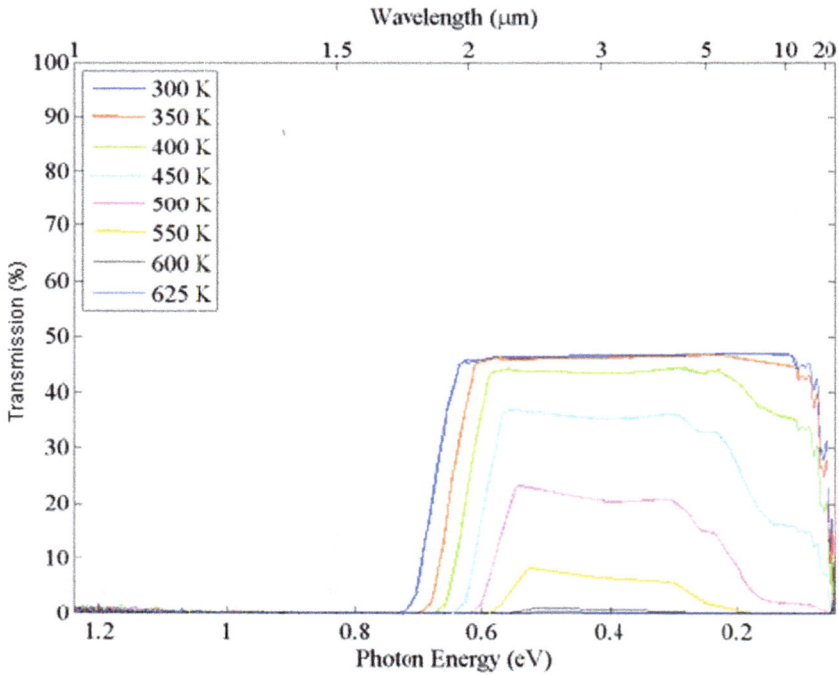

Figure 5.8. Mid-wave infrared transmission of Ge as a function of photon energy at various temperatures [6].

References

[1] http://www.ioffe.ru/SVA/NSM/Semicond/Ge/bandstr.html Accessed 10 November 2016
[2] Carruthers J A, Geballe T H, Rosenburg H M and Ziman J M 1957 The thermal conductivity of germanium and silicon between 2 and 300 degrees K *Proc. R. Soc.* A **238** 502–14
[3] http://www.ioffe.ru/SVA/NSM/Semicond/Ge/thermal.html Accessed 10 November 2016
[4] Ulmer E A Jr and Frankl Daniel R 1969 Infrared emission from free carriers in germanium *J. Opt. Soc. Am.* **59** 1233–9
[5] Moss T S and Hawkins T D H 1958 The infra-red emissivities of indium antimonide and germanium *Proc. Phys. Soc* **72** 270–3
[6] Harris T R 2010 Optical properties of Si, Ge, GaAs, GaSb, InAs and InP at elevated temperatures *Thesis* Department of the Air force, Air Force Institute of Technology

Chapter 6

Graphene

Graphene is a two-dimensional mono layer of hexagonally arranged carbon atoms. Graphene is considered as a zero gap semiconductor, since its valence and conduction bands meet at the Dirac points. Graphene has a high thermal conductivity due to large mean free path of phonons. Due to its high thermal conductivity, graphene has been used in various thermal management applications.

In general, the resistance of graphene decreases with increase in temperature. Around room temperature, its resistance decreases with decrease in temperature implying a positive TCR [1]. The low energy phonons in graphene contribute to its high thermal conductivity. Electron mobilities in graphene are weakly dependent on temperature.

Graphene nanostructures exhibit near black body radiation spectrum [2]. In figure 6.1, the black body radiation spectra of graphene are shown. Almost perfect blackbody radiation spectra with an emissivity of 0.99 can be observed.

The specific heat is given by the change in energy density when the change in temperature is 1 Kelvin. The terms specific heat and heat capacity may be used interchangeably. The specific heat shows how quickly the hot body cools or heats. Thermal time constants are usually very short for materials at the nano-scale. The time constant for single sheet graphene is 0.1 ns [3].

Figure 6.2 shows the specific heat of graphene, graphite, and diamond, all dominated by phonons at temperatures above ~1 K. The specific heat of graphene is dominated by phonons at all temperatures. At room temperature, the thermal conductivity of graphene/SiO_2 is measured to be ~600 W m^{-1} K^{-1}, and that of SiO_2-encased graphene 40 is ~160 W m^{-1} K^{-1} [3].

Figure 6.3 shows the emissivity of graphene versus wavelength for few layers of graphene at room temperature. The emissivity is almost constant in the wavelength range of 1–2 μm at room temperature. The 2.3% IR absorption in a single layer of

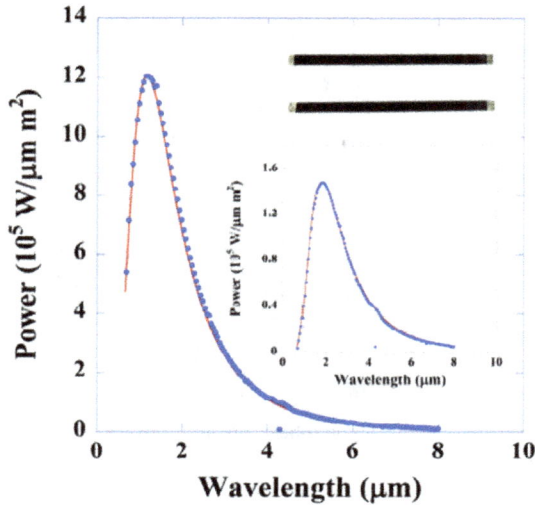

Figure 6.1. Blackbody radiation spectra of graphene nanostructure at a temperature of 2480 K. The upper inset displays the radiation spectrum of graphene nanostructure before and after heating to 2600 K. The lower inset shows the radiation spectrum of graphene nanostructure at 1600 K [2].

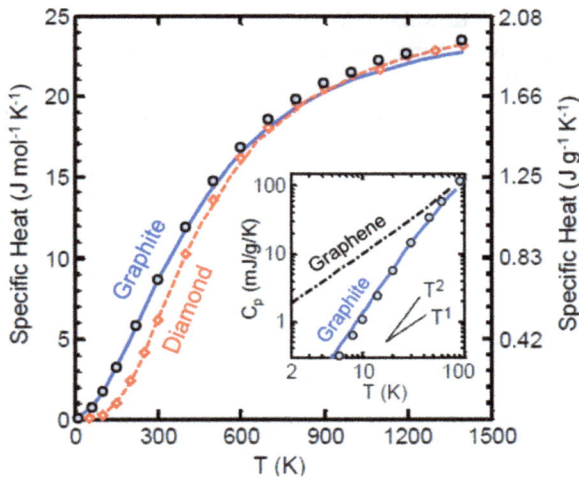

Figure 6.2. Specific heat of graphene, graphite, and diamond, all dominated by phonons at temperatures above ~1 K.

graphene is attributed to inter-band absorption in a wide range of wavelengths. The emissivity increases by about 20% with increment of each layer. Graphene has a very low reflectivity, and most of the incident electromagnetic waves are found to be transmitted (~97%) [4].

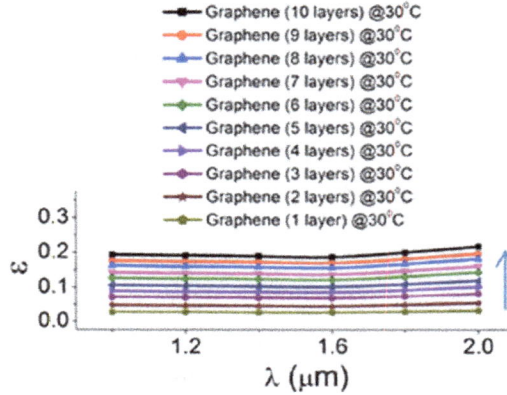

Figure 6.3. Emissivity versus wavelength for graphene up to 10 layers at room temperature. Reproduced from [4] with permission of Springer.

References

[1] Obeng Y and Sirinivasan P 2011 Graphene: is it the future for semiconductors? An overview of the material, devices, and applications, The Electrochemical Society Interface, Spring 2011 47–52

[2] Matsumoto T, Koizumi T, Kawakami Y, Okamoto K and Tomita M 2013 Perfect blackbody radiation from a graphene nanostructure with application to high temperature spectral emissivity measurements *Opt. Express* **21** 30964–74

[3] Pop E, Varshney V and Roy A K 2012 Thermal properties of graphene: fundamentals and applications *MRS Bull.* **37** 1273–81

[4] Muley S V and Ravindra N M 2014 Emissivity of electronic materials, coatings, and structures *JOM* **66** 616–35

Chapter 7

Silicon carbide

Silicon carbide (SiC), also commonly known as carborundum, is a compound of silicon and carbon. Apart from being an excellent abrasive, SiC is also used in the manufacture of semiconductor process equipment. SiC is a very hard material and is thermally stable up to 1600 °C.

Carrier concentration is a function of thermal energy. The population density of electrons, holes or phonons is sensitive to temperature. SiC has a high thermal conductivity, making it an appropriate candidate for high temperature applications. The thermal conductivity of single crystal SiC is 500 W m^{-1} K^{-1} [1]. The thermal conductivity of SiC may be calculated by using the expression [1]:

$$\kappa = \alpha \cdot \rho \cdot C_{\mathrm{p}} \tag{7.1}$$

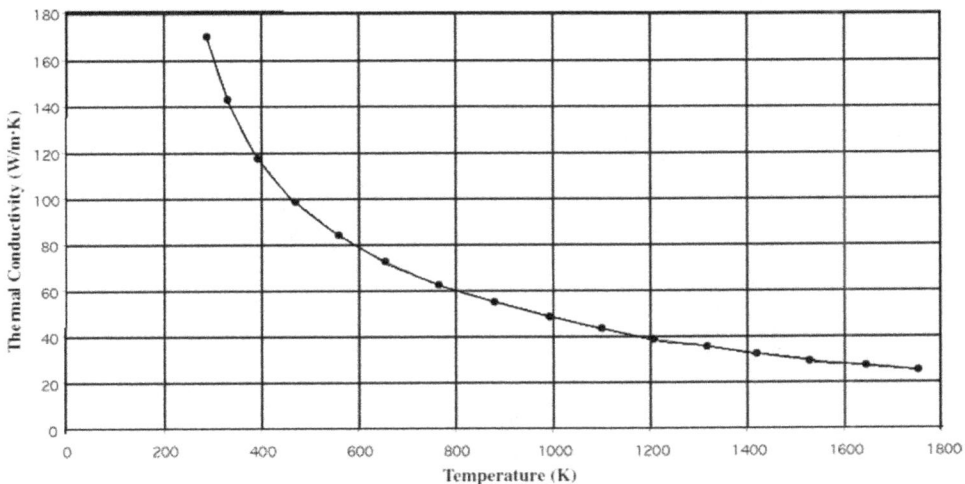

Figure 7.1. Thermal conductivity of super pure SiC as a function of temperature [1]. Reproduced with the permisison of POCO Graphite Inc.

doi:10.1088/978-1-6817-4112-3ch7

where κ is thermal conductivity, α is thermal diffusivity, ρ is bulk density and C_p is heat capacity.

The heat capacity of SiC is calculated by the equation [1]:

$$C_p, \text{SiC} = 1.267 + 0.049*10^{-3}\, T - 1.227*10^5\, T^{-2} + 0.205*10^8 T^{-3} \qquad (7.2)$$

where T is absolute temperature.

Figure 7.1 shows the thermal conductivity as a function of temperature for super pure silicon carbide. Figure 7.2 shows the thermal conductivity of highly pure SiC and some less pure SiC samples.

Representative values of the thermal conductivity of highly pure SiC samples are summarized in table 7.1.

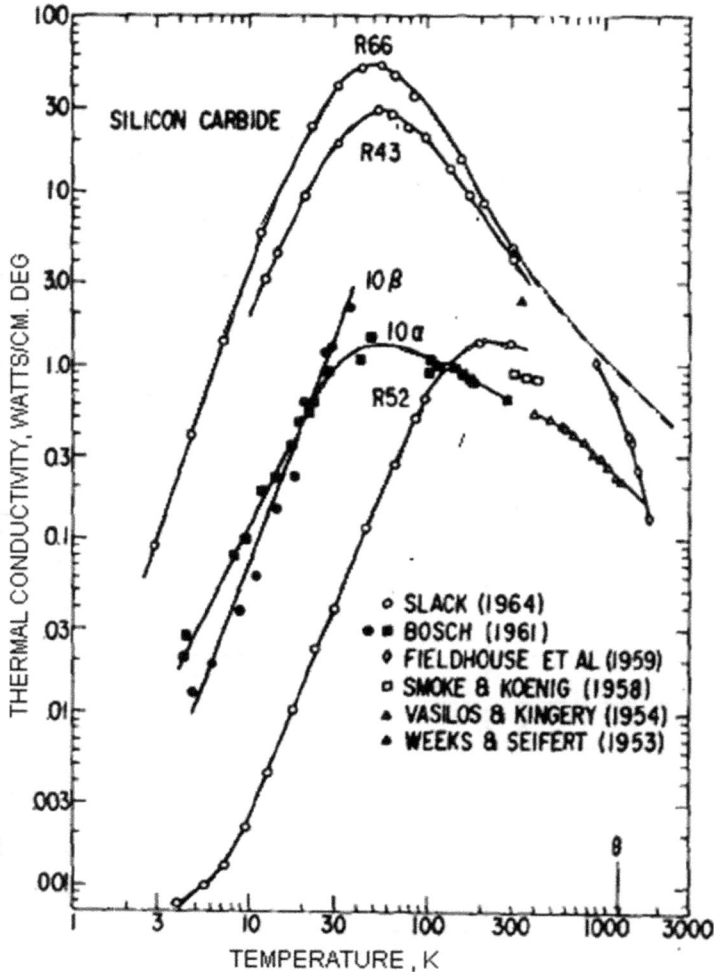

Figure 7.2. Thermal conductivity κ versus temperature of high purity SiC (R66) and various less pure samples. Reprinted from [2] with the permission of AIP Publishing.

Table 7.1. Representative values of thermal conductivity of high-purity SiC sample R66 [2].

T (K)	κ (W cm^{-1} deg^{-1})
50	52
70	43
100	28
150	15
200	9.5
300	4.9
500	2.4*
1000	1.1*
1200	0.9*

*= extrapolated.

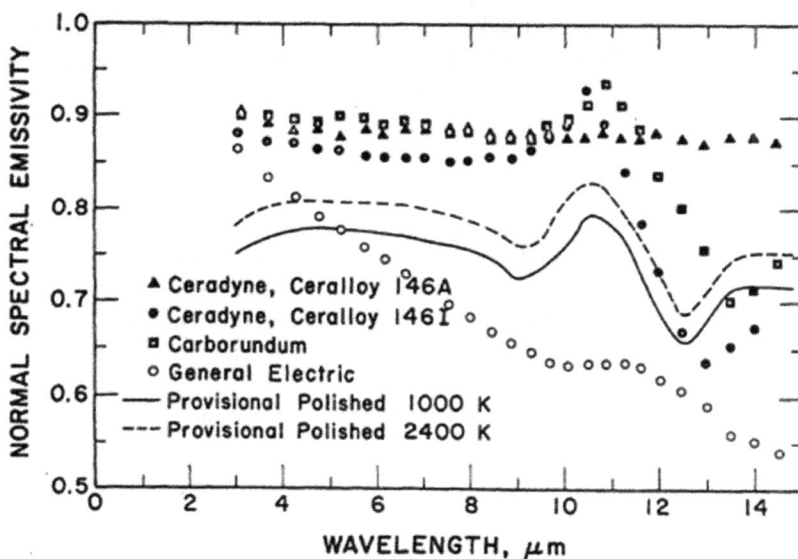

Figure 7.3. Normal spectral emissivity of SiC at 1900 K [3].

   In pure SiC, the heat transport is a result of phonons. The thermal conductivity in impure SiC is mainly contributed by phonon scattering from charge carriers that are associated with impurities in materials [2].

   Figure 7.3 shows the normal spectral emissivity of SiC at 1900 K. The spectral emissivity of SiC may be controlled at longer wavelengths for specific applications [3]. The following features can be observed in the figure: a dip in emissivity (0.726) at 9 µm followed by a peak in emissivity (0.795) at 10.5 µm. A subsequent dip in emissivity (0.658) occurs at 12.5 µm. The emissivity (0.716) remains constant at wavelengths beyond 13.6 µm.

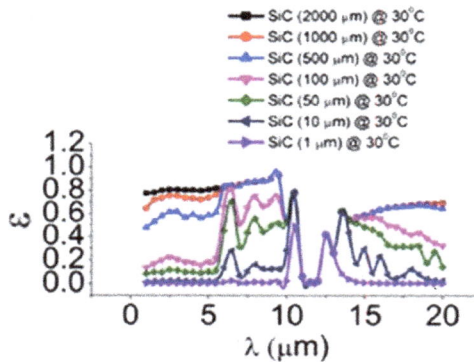

Figure 7.4. Emissivity of SiC as a function of wavelength for different thicknesses. Reproduced from [4] with permission of Springer.

From figure 7.4, it can be observed that the emissivity of SiC increases with thickness. For samples with thickness $\geqslant 500$ μm, the emissivity increases with wavelength from 0 to 10 μm beyond which the trend is complex. The emissivity reaches values up to ~0.8–0.98. The emissivity becomes negligible at the wavelength of 11.6 μm.

References

[1] Rashed A H 2002 Properties and characteristics of silicon carbide, POCO Carbide Inc.
[2] Slack G A 1964 Thermal conductivity of pure and impure silicon, silicon carbide, and diamond *J. Appl. Phys.* **35** 3460–6
[3] Taylor R E, DeWitt D P and Johnson P E 1980 Spectral emissivity at high temperatures, Thermophysical Properties Research Laboratory, Annual report for AFOSR grant 77-3280 pp 1–14
[4] Muley S V and Ravindra N M 2014 Emissivity of electronic materials, coatings, and structures *JOM* **66** 616–35

Chapter 8

Gallium arsenide

Gallium arsenide (GaAs) is a III–V compound direct bandgap semiconductor. GaAs is used in the manufacture of optoelectronic devices such as solid state lasers, light emitting diodes (LEDs), solar cells etc. It is also used as a substrate material for the epitaxial growth of other III–V compound semiconductors including indium gallium arsenide (InGaAs), aluminum gallium arsenide (AlGaAs) etc.

GaAs has a thermal conductivity of 0.55 W cm^{-1} °C^{-1}, which is about one-third that of silicon and one-tenth that of copper. The reliability of GaAs devices is directly related to the thermal characteristics of the device design. The thermal conductivity of GaAs is related to the temperature of the material over a wide temperature range and varies approximately as $1/T$, where T is the temperature in Kelvin. While GaAs has a higher thermal resistivity than silicon, it is somewhat offset by the higher band gap of GaAs, allowing higher operating temperatures.

In semiconductors such as GaAs, electronic transition mechanisms dominate in the visible wavelength region and optical phonon absorption dominates in the far IR range of wavelengths.

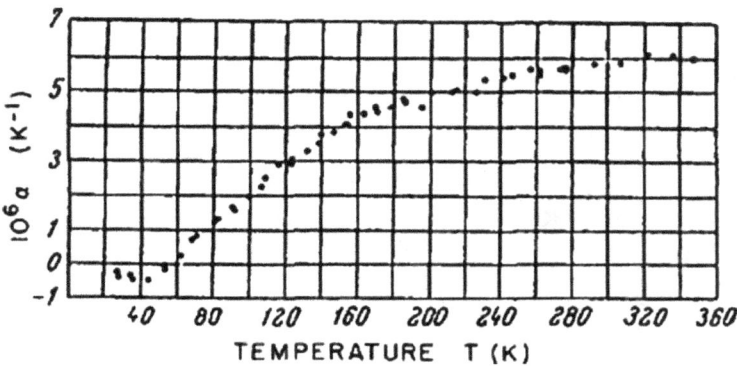

Figure 8.1. Linear expansion coefficient for GaAs in the temperature range of 28–348 K. Reprinted from [1] with the permission of AIP Publishing.

doi:10.1088/978-1-6817-4112-3ch8
8-1

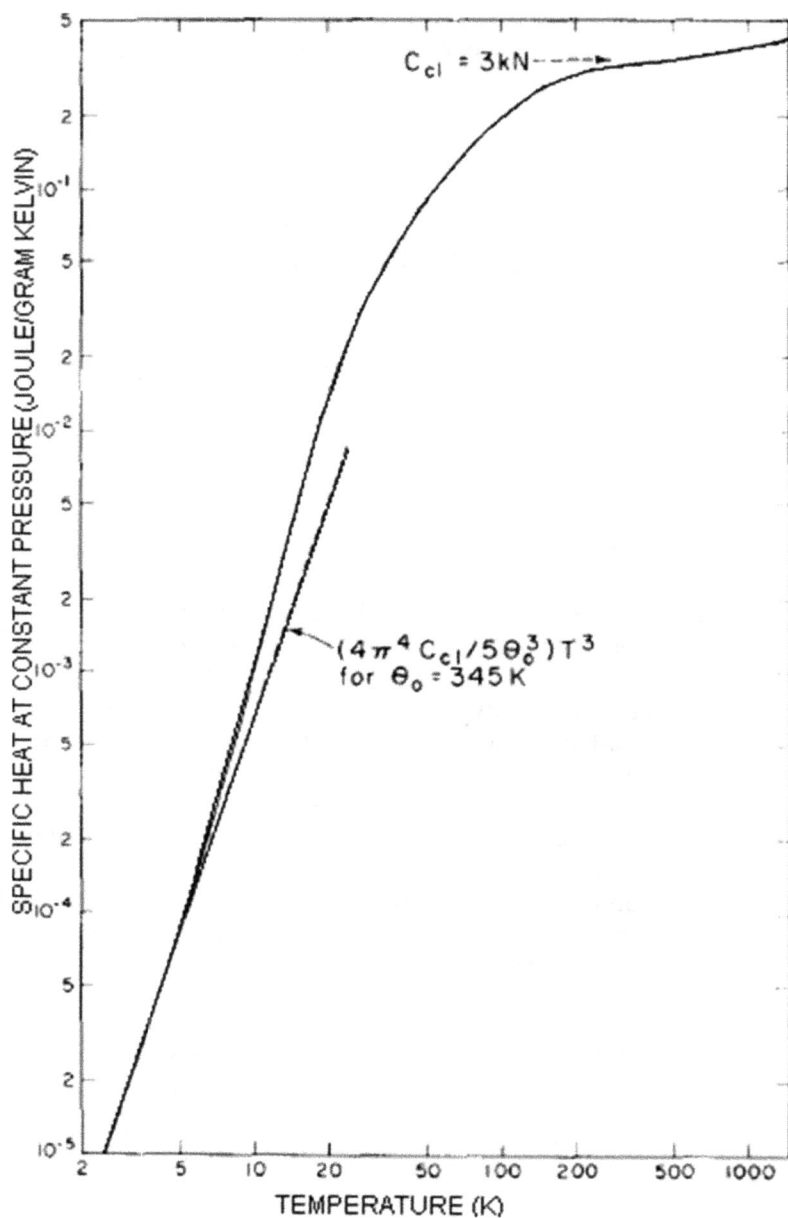

Figure 8.2. Specific heat from 3 K to ~1000 K. Reprinted from [1] with the permission of AIP Publishing.

The linear expansion coefficient (α) of GaAs decreases from its high temperature positive value on cooling, passes through zero for ~56 K [1]. α, as a function of temperature, in the range of 28–348 K is shown in figure 8.1.

The pressure dependences of elastic moduli for diamond-type lattices, such as GaAs, should be consistent with a positive expansivity, $\alpha_T \propto +T^3$, for $T \ll \theta_D$.

The 'classical' specific heat at constant volume for N atoms of solids ($= \frac{1}{2}N$ molecules for diatomic GaAs) is $C = 3kN$. Specific heat at constant pressure, C_p, for temperatures up to Debye temperature (θ_D) is plotted in figure 8.2.

GaAs exhibits isotropic thermal conduction by transport of phonons, photons or electrons or holes. Figure 8.3 shows the low temperature lattice thermal conductivity κ_L for four variously doped GaAs monocrystalline samples [1]. The temperature dependence of κ_L for the various samples is appreciably steeper than the T^{-1} dependence.

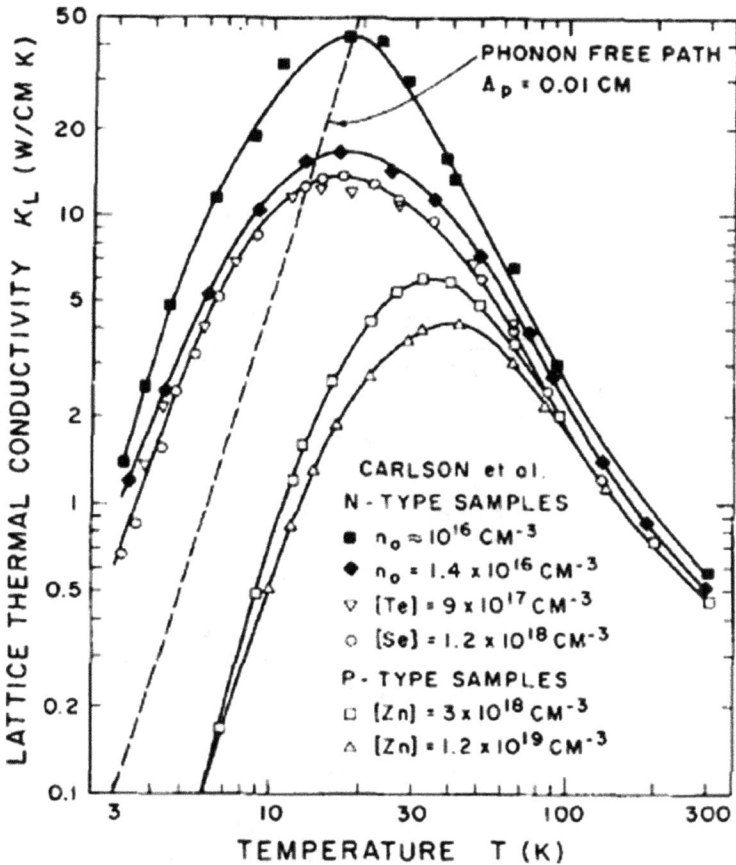

Figure 8.3. Lattice thermal conductivity from 3 to 300 K for various GaAs samples with various doping concentrations. Reprinted from [1] with the permission of AIP Publishing.

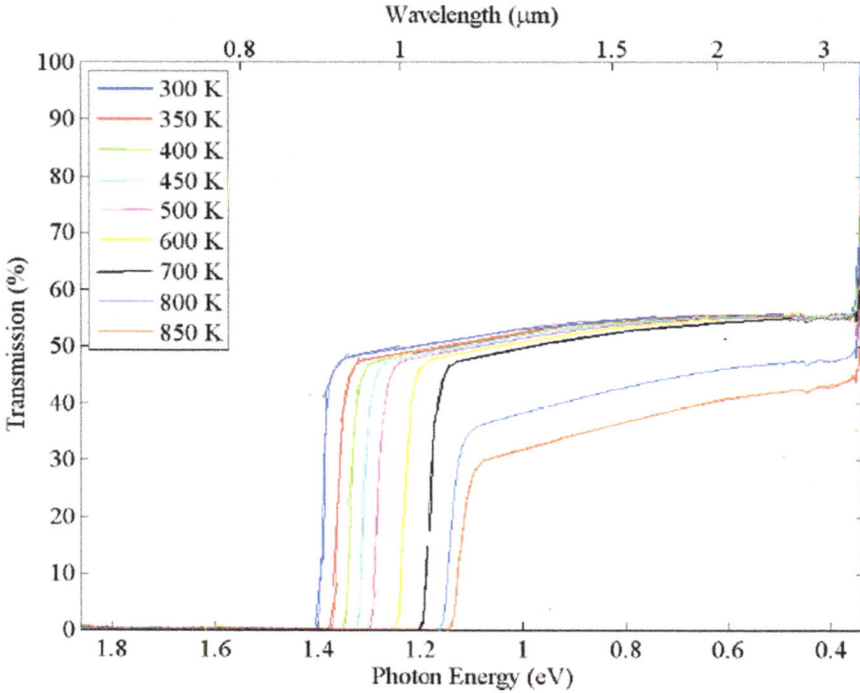

Figure 8.4. Near infrared transmission versus photon energy for GaAs at several temperatures [2].

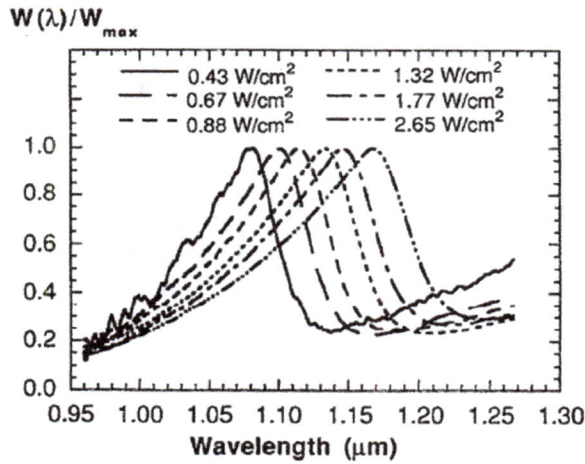

Figure 8.5. Normalized emission spectra from a GaAs specimen heated with a range of electron beam power densities. Reprinted from [3] with the permission of AIP Publishing.

Temperature dependent transmission measurements of GaAs are shown in figure 8.4. The absorption edge shifts to lower energy (to the right) as temperature increases. Also, a decrease in the overall transmission can be observed for photon energies less than ~1.1 eV starting at about 800 K [2].

Total Hemispherical Emissivity

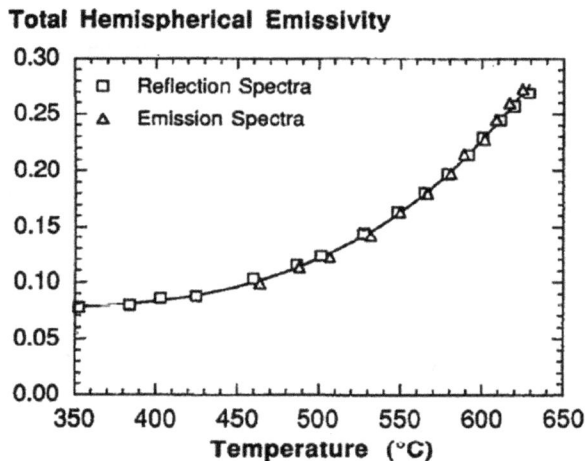

Figure 8.6. Total hemispherical emissivity of GaAs as a function of temperature. Reprinted from [3] with the permission of AIP Publishing.

GaAs samples of 485 μm thickness have been considered by Timans to study the normalized spectral emissivity [3]. Figure 8.5 shows the normalized emission spectra recorded from the specimen heated with a range of power densities. The decrease in the band gap of GaAs with temperature can be observed by the falling edge of the spectrum at longer wavelengths as heating power is increased.

The spectral emissivity for wavelengths beyond the absorption edge of GaAs is about 10% of the emissivity at wavelengths where the specimen is opaque.

Figure 8.6 shows the total hemispherical emissivity of GaAs as a function of temperature. A significant increase in the total emissivity at temperatures above 500 °C can be seen in figure 8.6. This increase in emissivity is due to free carrier absorption. It becomes more pronounced with increase in temperature beyond 500 °C. The small total emissivity, at low temperatures, makes it easy to attain high equilibrium temperatures for low heating power densities. The increase in emissivity is almost doubled between 500 °C and 600 °C.

References

[1] Blakemore J S 1982 Semiconducting and other major properties of gallium arsenide *J. Appl. Phys.* **53** R123–81
[2] Harris T R 2010 Optical properties of Si, Ge, GaAs, GaSb, InAs and InP at elevated temperatures *Thesis* Department of the Air force, Air Force Institute of Technology
[3] Timans P J 1992 The experimental determination of the temperature dependence of the total emissivity of GaAs using a new temperature measurement technique *J. Appl. Phys.* **72** 660–70

Radiative Properties of Semiconductors

N M Ravindra, Sita Rajyalaxmi Marthi and Asahel Bañobre

Chapter 9

Gallium nitride

Gallium nitride (GaN) is considered as one of the most important semiconductors after silicon. It is a direct bandgap III–V compound semiconductor. With a wide band gap of 3.4 eV, GaN finds several significant applications in the optoelectronics industry including the blue LEDs. Additionally, GaN based transistors can operate at high temperatures. Another advantage of GaN is that it has low electrical resistance implying low power loss via heat.

Thermal property measurements made on GaN [1], in the temperature range of 300–900 K, show that the thermal expansion coefficient α is 5.59×10^{-6} K^{-1}. In the temperature ranges of 300–700 K and 700–900 K, the values of α are 3.17×10^{-6} K^{-1} and 7.75×10^{-6} K^{-1}, respectively. The thermal expansion coefficient depends on various factors such as defect concentration, free charge carrier concentration and strain in the crystal lattice. Linear expansion coefficient, as a function of temperature, of GaN is shown in figure 9.1.

The temperature dependence of thermal conductivity in GaN depends mainly on intrinsic phonon–phonon scattering. The thermal conductivity of GaN is lower when compared to other semiconductors. The temperature dependence of thermal conductivity of free standing GaN is shown in figure 9.2.

The temperature dependence of specific heat of GaN at constant pressure (C_p), in the temperature range of 298 K $< T <$ 1773 K, is expressed by [1]:

$$C_p(T) = 9.1 + (2.14 \times 10^{-3}\, T)\ (\text{J mol}^{-1}\,\text{K}^{-1}), \tag{9.1}$$

$$C_p = 38.1 + 8.96 \times 10^{-3}\, T\ (\text{cal mol K}^{-1}), \tag{9.2}$$

doi:10.1088/978-1-6817-4112-3ch9

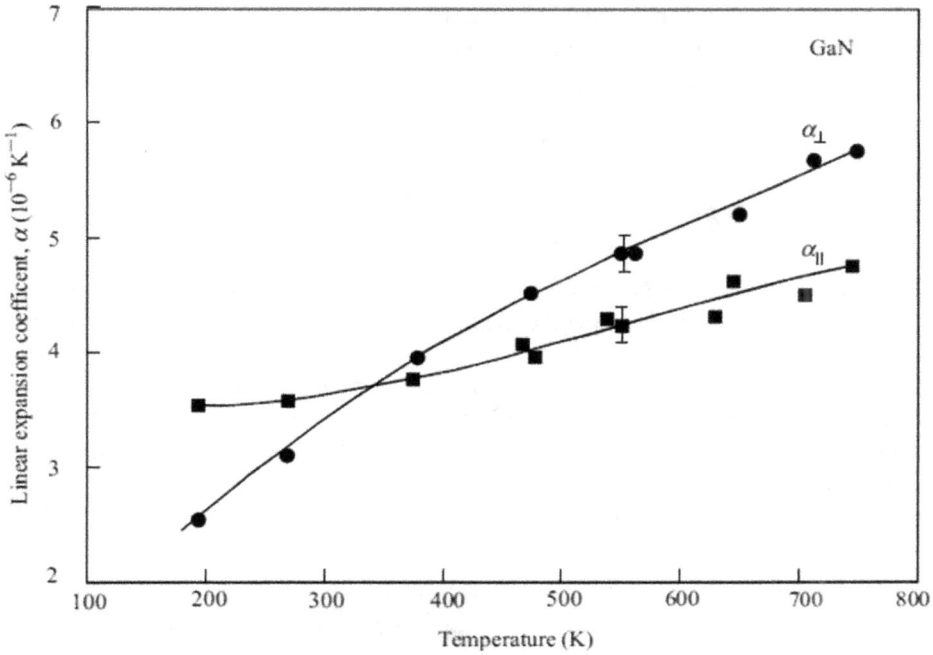

Figure 9.1. Coefficient of linear thermal expansion of GaN versus temperature. Reprinted from [1] with permission of John Wiley & Sons.

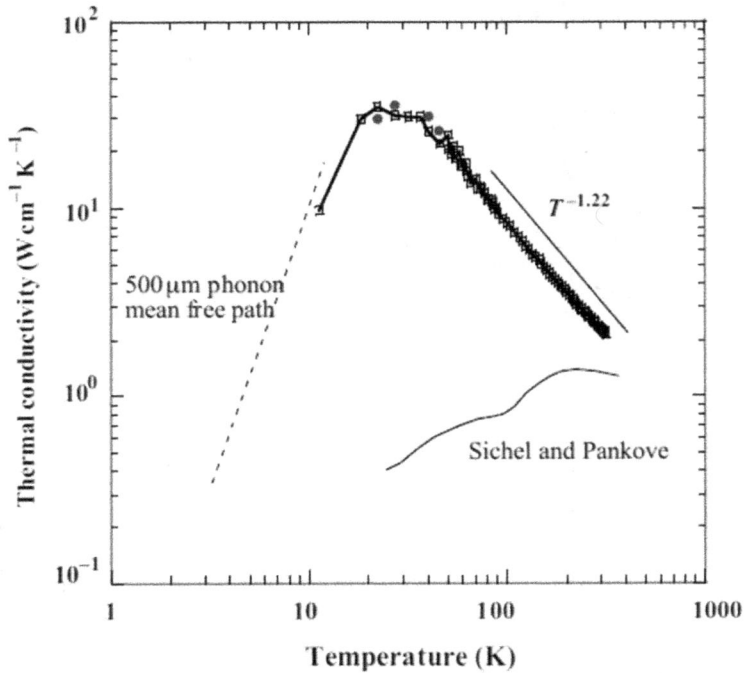

Figure 9.2. Thermal conductivity of 200 μm free standing GaN sample as a function of temperature. Reprinted from [1] with permission of John Wiley & Sons.

The Debye expression for the temperature dependence of specific heat in a solid at a constant pressure (C_p) can be expressed as:

$$C_p = 18R\left(\frac{T}{\theta_D}\right)^3 \cdot \int_0^{x_D} \frac{x^4 e^x}{(e^x - 1)^2}dx,$$ (9.3)

where, $x_D = \theta_D/T$,

$$R = 8.314\,\text{J mol}^{-1}\,\text{K}^{-1}$$

The specific heat of GaN at constant pressure has T^3 dependence like most semiconductors. The molar specific heat at constant pressure, as a function of temperature, is shown in figure 9.3.

The summary of thermal properties of GaN is presented in table 9.1.

Figure 9.4 shows the emissivity of GaN as a function of wavelength for various thicknesses. The emissivity of GaN increases with increase in thickness in the wavelength range of 3–10 μm. For thicknesses beyond 5 μm, the emissivity of GaN saturates at a value of ~0.89. A constant increase in emissivity is observed for all wavelengths beyond 3 μm.

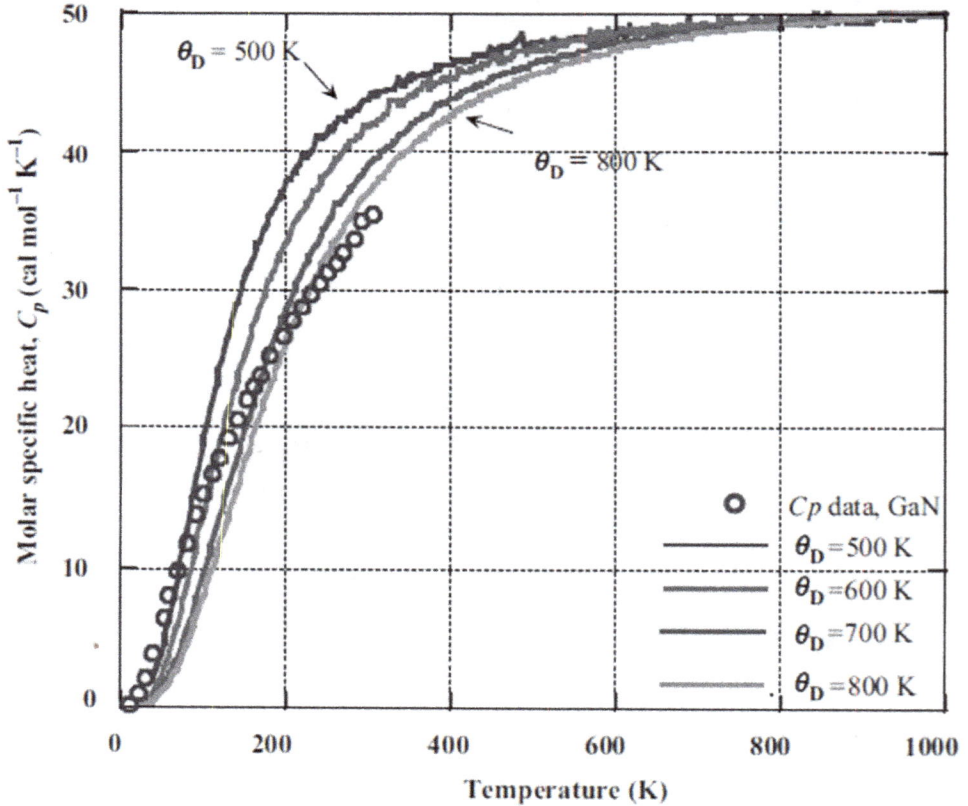

Figure 9.3. Molar specific heat at constant pressure of GaN versus temperature. Reprinted from [1] with permission of John Wiley & Sons.

9-3

Table 9.1. Parameters related to thermal properties of gallium nitride [1].

GaN	Parameter value/comments
Temperature coefficient (eV K^{-1})	$dE_g/dT = -6.0 \times 10^{-4}$
Thermal expansion (K^{-1})	$\Delta a/a = 5.59 \times 10^{-6}$, $\alpha_\parallel = \alpha_a = 5.59 \times 10^{-6}$ (wurtzite structure)
Thermal conductivity κ (W cm^{-1} K^{-1})	11.9 at 77 K, 2.3 at 300 K, 1.5 at 400 K
Debye temperature (K)	600
Melting point (°C)	>1700 (at 2 kbar), 2500 (at tens of kbar)
Specific heat (J g^{-1} °C^{-1})	0.49
Thermal diffuslvity (cm^2 s^{-1})	0.43
Heat of formation, ΔH_{298} (kcal mol^{-1})	−26.4
Heat of itemization, ΔH_{298} (kcal mol^{-1})	−203
Heat of sublimation (kcal mol^{-1})	72.4 ± 0.5
Heat capacity (J mol^{-1} K^{-1})	35.4 at 300 K
Specific heat (J mol^{-1} K^{-1}) (298 K < T < 1773 K)	$C_p = 38.1 + 8.96 \times 10^{-3}T$
Enthalpy, ΔH^0 (kcal mol^{-1})	−37.7
Standard entropy of formation, ΔS^0 (cal mol^{-1} K^{-1})	−32.43

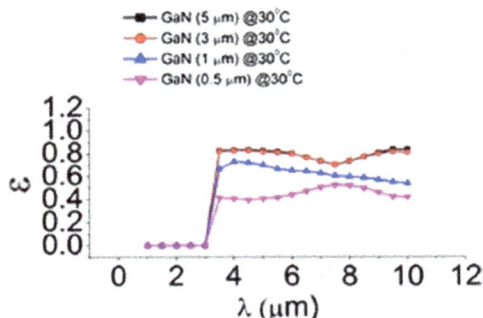

Figure 9.4. Emissivity of GaN as a function of wavelength for various thicknesses. Reproduced from [2] with permission of Springer.

References

[1] Morkoç H General properties of nitrides, Handbook of Nitride Semiconductors and Devices: Materials Properties, Physics and Growth vol 1 (New York: Wiley)
[2] Muley S V and Ravindra N M 2014 Emissivity of electronic materials, coatings, and structures *JOM* **66** 616–35

Chapter 10

Indium antimonide

Indium antimonide (InSb) is a III–V compound semiconductor. It has several applications as infrared detectors, thermal imaging cameras etc. InSb detectors are sensitive in the wavelength range of 1–5 μm.

Among all the III–V semiconducting compounds, InSb has the lowest melting temperature at 523 °C [1].

The energy gaps at 300 K, 100 K and 0 K are usually taken to be about 0.17, 0.22 and 0.23 eV, respectively [2]. The number of intrinsic carriers at 300 K is 2×10^{16} cm^{-3}. The electron and hole mobilities in the highest purity material are, respectively, about 7×10^4 and 7×10^2 cm^2 (V-s)$^{-1}$ at 300 K and about 6×10^5 and 1×10^4 cm^2 (V-s)$^{-1}$ at 77 K. The low energy gap in InSb leads to a very high absorption coefficient associated with band-to-band transitions for wavelengths below 5–7 μm [2]. The material may therefore be used as infrared detector and filter.

The linear expansion coefficient of InSb is 6.50×10^{-6} per °C and 5.04×10^{-6} per °C at 80 K and 300 K, respectively; below 80 K, it decreases, even becoming slightly negative, before approaching zero at low temperatures ~0 K; above 300 K, it increases slightly with respect to the 300 K value, reaching about 5.4×10^{-6} per °C at 200 °C, but it decreases to zero at about 500 °C [2].

Table 10.1. Electrical properties of InSb at various temperatures [3].

T (K)	ρ (Ω-cm)	μ (cm^2 V-s^{-1})	N (cm^{-3})
83	6×10^{-3}	(173 600)	6×10^{15}
217	10×10^{-3}	(104 160)	6×10^{15}
300	4×10^{-3}	78 000	2×10^{16}
400		44 000	8×10^{16}
500		28 500	

doi:10.1088/978-1-6817-4112-3ch10

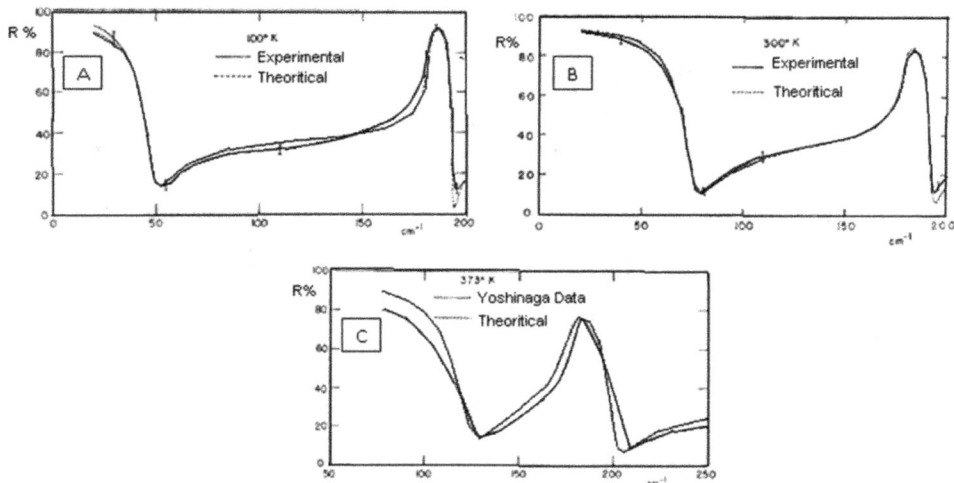

Figure 10.1. Reflectivity of InSb at 100 K, 300 K and 373 K, theoretically and experimentally [3].

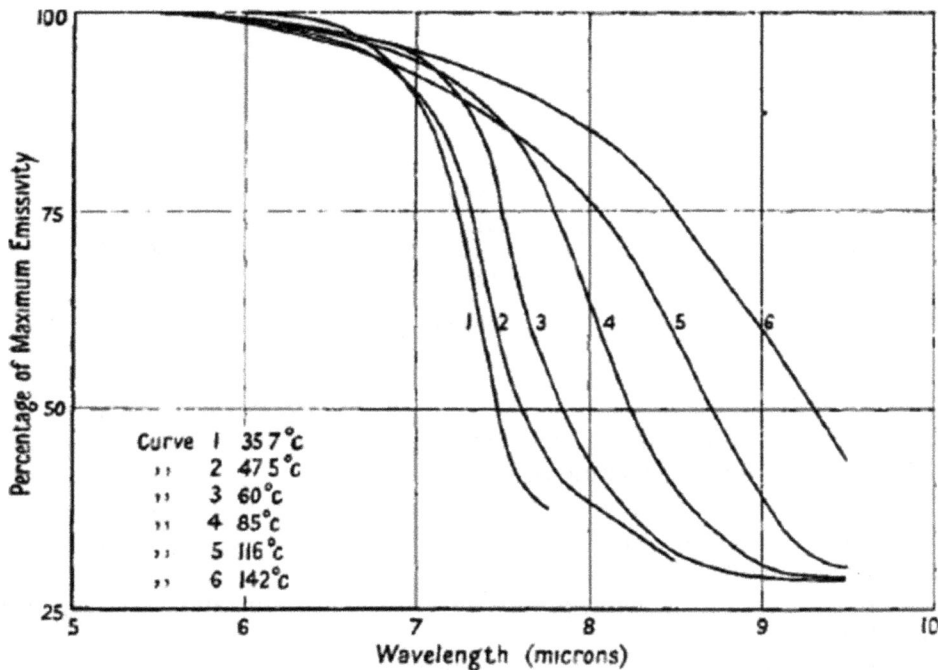

Figure 10.2. Emissivity of InSb at various temperatures. Reprinted from [4] with permission from Elsevier.

The thermal conductivity has been measured to be 0.04 at 400 °C and 0.02 at 425 °C in cal (cm s)$^{-1}$.

Various electrical properties of InSb, as a function of temperature, are shown in table 10.1.

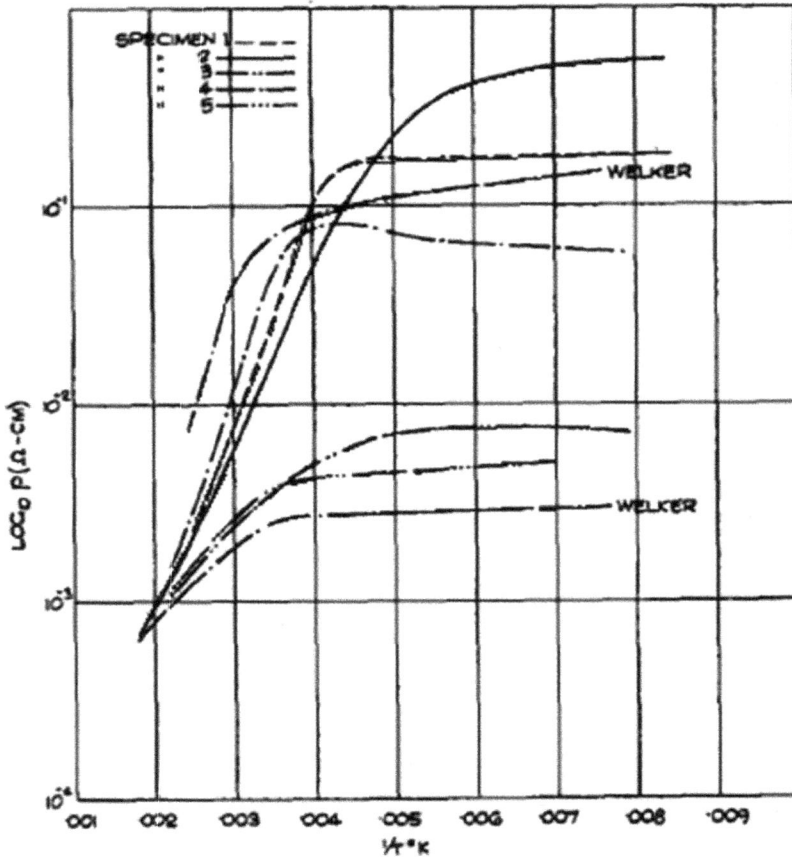

Figure 10.3. Resistivity of InSb as a function of absolute temperature. Reproduced with permission from [5].

In the far infrared region, the optical properties are dependent on the conduction band electrons and lattice vibrations.

Reflectivity of InSb, at various temperatures, is shown in figure 10.1. At 100 K, the free carrier reflectivity has a minimum of 14% at 53.5 cm^{-1} (~ 186 µm). At 300 K, it becomes 10.4% at 78.5 cm^{-1} (~ 127 µm).

The plots of relative emissivity of InSb at various temperatures are shown in figure 10.2.

The emissivity of InSb is high at wavelengths below band edge and has low emissivity at long wavelengths. This behavior is expected in semiconductors. It may be observed that the absorption edge moves to longer wavelengths for elevated temperatures.

The electrical resistivity and conductivity of InSb, as a function of temperature, are plotted in figures 10.3 and 10.4, respectively. These figures, taken from different sources in the literature, facilitate a comparison of its electrical properties.

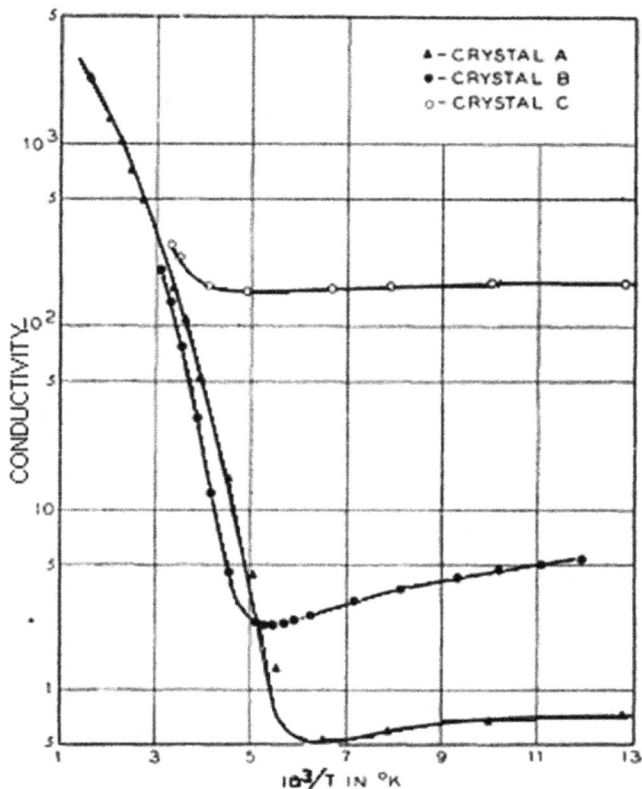

Figure 10.4. Electrical conductivity of InSb as a function of reciprocal temperature. Reproduced from [6] with permission from Elsevier.

References

[1] Welker H 1954 Semiconducting Intermetallic Compounds *Physica* **20** 893–909

[2] Hulme K F and Mullin J B 1962 Indium antimonide—a review of its preparation, properties and device applications *Solid-State Electron* **5** 211–47

[3] Sanderson R B 1965 Far infrared optical properties of indium antimonide *J. Phys. Chem. Solids* **26** 833–910

[4] Moss T S and Hawkins T D H 1958 The infra-red emissivities of indium antimonide and germanium *Proc. Phys. Soc.* **72** 270–3

[5] Austin I G and Mcclymont D R 1954 The interpretation of Hall effect, conductivity and infrared measurements in indium antimonide *Physica* **20** 1077–83

[6] Tanenbaum M and Maita J P 1953 Hall effect and conductivity of InSb single crystals *Phys. Rev.* **91** 1009–10

Chapter 11

Indium phosphide

Indium phosphide (InP) is a III–V compound direct bandgap semiconductor. InP has several applications in optoelectronics and high frequency electronic devices. InP is an attractive alternative to GaAs for the following reasons:

- Higher thermal conductivity implying less sensitivity to temperature in device performance parameters
- Reliable circuits due to superior surface passivation capabilities
- Improved thermal properties, making it a suitable candidate for devices with higher current densities
- Increased electron mobility when compared to GaAs.

The thermal conductivity of InP [1] is 0.68 Wcm^{-1} K^{-1}.

The optical properties of two InP samples, one Fe-doped, 550 μm thick highly resistive and the other S-doped, 450 μm thick are presented in this chapter. The temperature dependent optical constants of S-doped InP samples are shown in figures 11.1 and 11.2.

In figure 11.1, for S-doped InP, the refractive index decreases with increase in wavelength, attains minima at ~17 μm and subsequently increases. While these values are generally lower than the data reported in the literature, at room temperature, it must be noted that the temperature dependence of these properties is generally scarcely available in the literature. Additionally, these properties are sensitive to doping, surface conditions etc.

The radiative properties of an Fe-doped InP sample at temperatures of 51, 234, 265, 374 and 479 °C, while heating the sample, and at a temperature of 40 °C after cooling the sample are shown in figures 11.3–11.5.

For Fe-doped InP sample, the emissivity is very low and almost constant in the wavelength range of 1 to 10 μm. At 14 μm, a sharp peak in the emissivity is observed due to multi-phonon absorption. The steep increase in the emissivity of InP at 479 °C at 1.2 μm is due to band-gap absorption.

doi:10.1088/978-1-6817-4112-3ch11   11-1

InP S-Doped

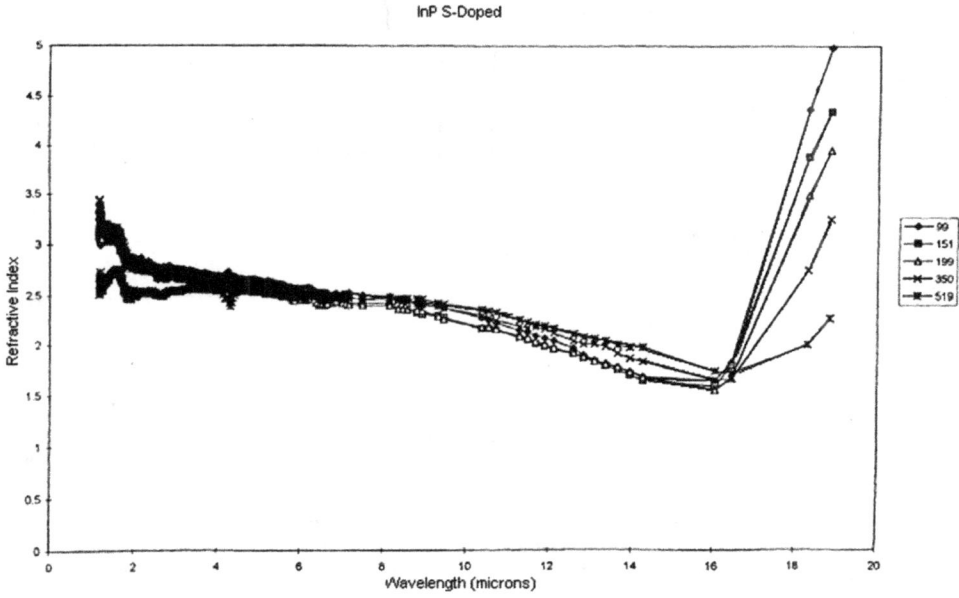

Figure 11.1. Refractive index versus wavelength plot of InP, S-doped at different temperatures [1].

InP S-Doped

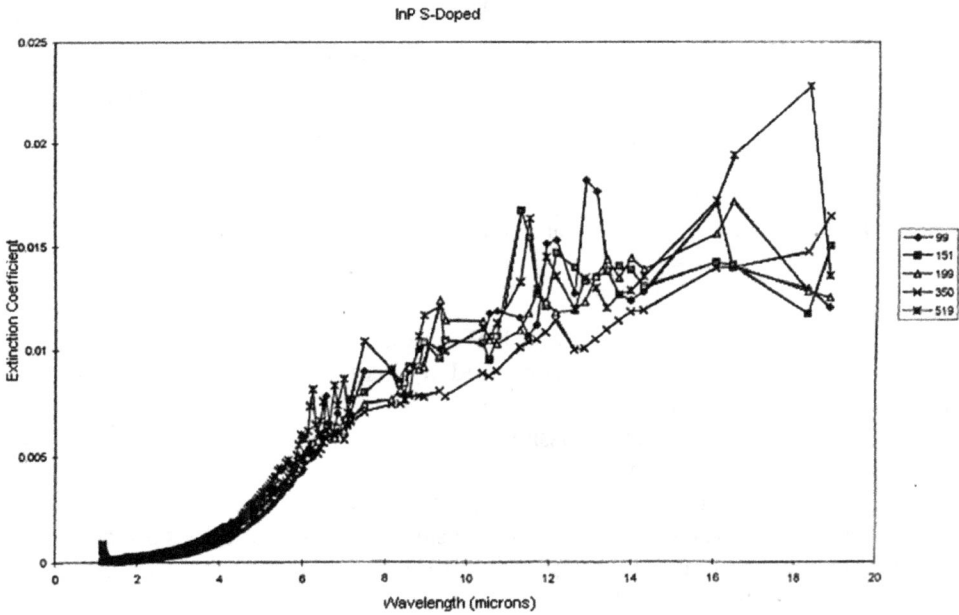

Figure 11.2. Extinction coefficient versus wavelength plot of InP, S-doped at different temperatures [1].

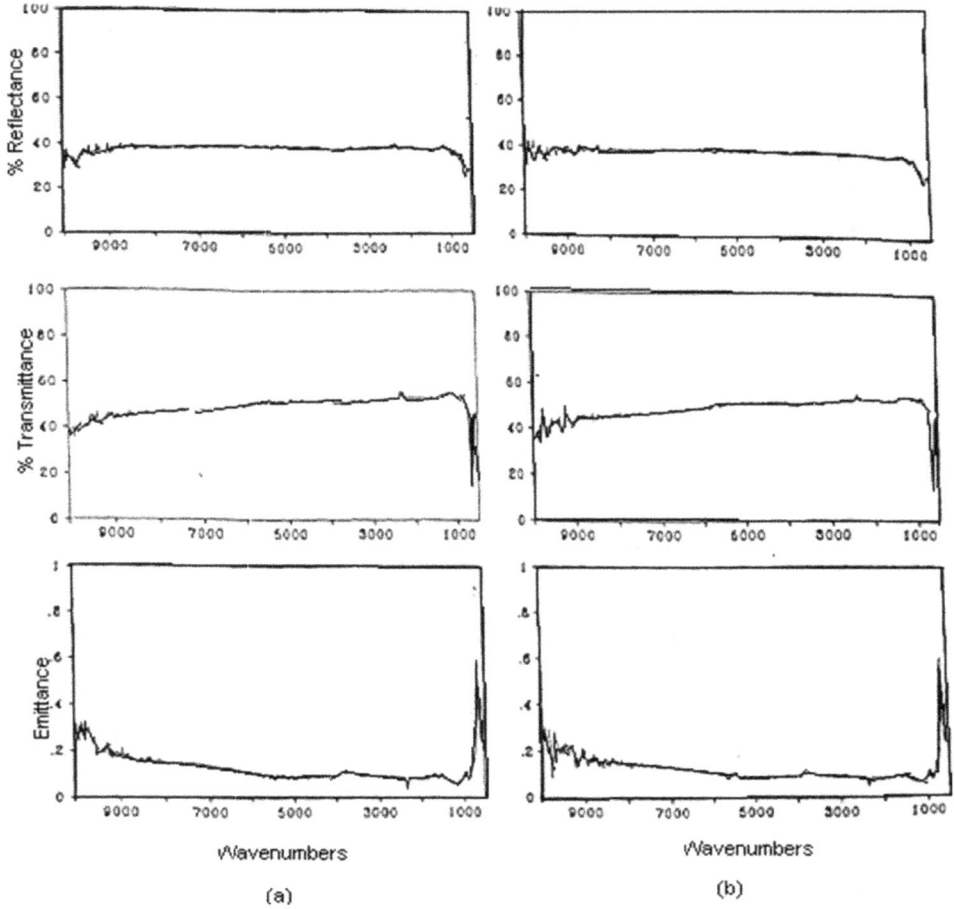

Figure 11.3. Temperature dependent optical properties of Fe-doped InP: (a) 234 °C and (b) 51 °C [1].

The radiative properties of S-doped InP sample at temperatures of 99, 199, 350, 421 and 519 °C while heating the sample and at a temperature of 92 °C after cooling the sample are shown in figures 11.6–11.8.

For S-doped InP sample, the emissivity increases as the temperature of the sample increases and reaches high values at 519 °C.

In S-doped InP sample, the dopant, S, contributes to the increase in free carriers of InP. However, in the case of Fe-doped InP sample, the dopant does not contribute to the free carriers. It may be anticipated that the emissivity values of Fe-doped InP are close to those of intrinsic InP.

The relation for the temperature dependence of specific heat at constant pressure is given by [2]:

$$C_p = 0.28 + 10^{-4} \cdot T \ (\text{J g}^{-1} \text{ K}^{-1}) \text{ for } 298 \text{ K} < T < 910 \text{ K} \qquad (11.1)$$

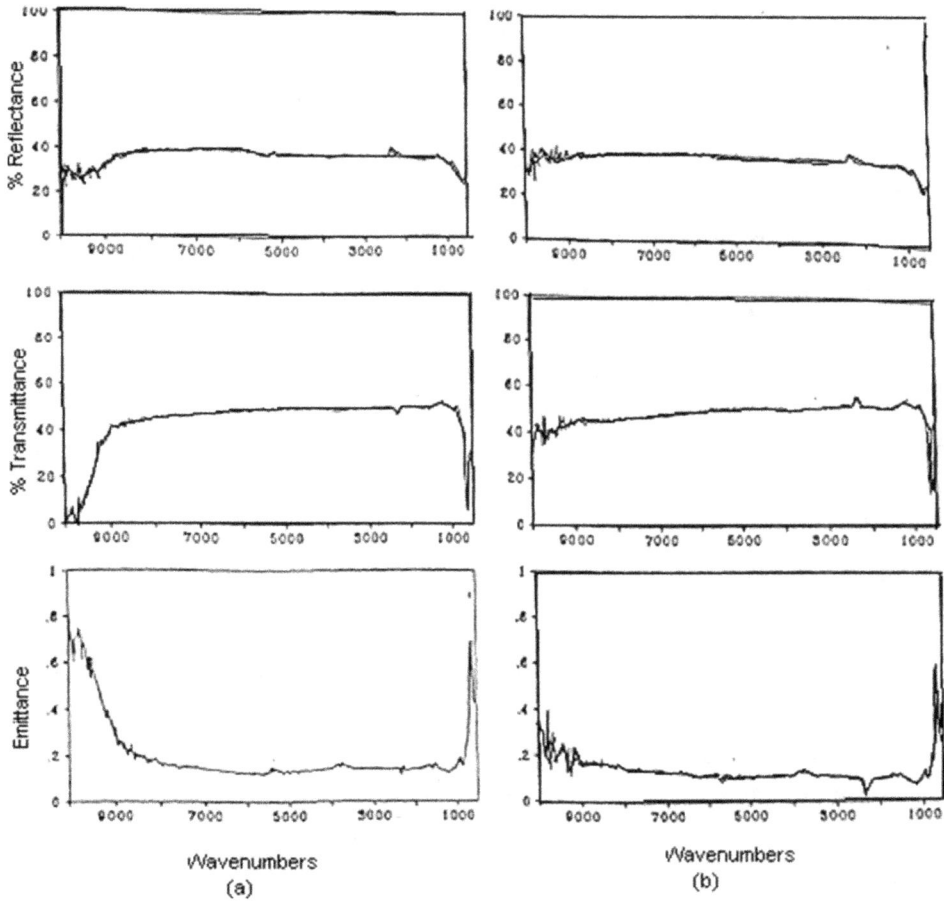

Figure 11.4. Temperature dependent optical properties of Fe-doped InP: (a) 374 °C and (b) 295 °C [1].

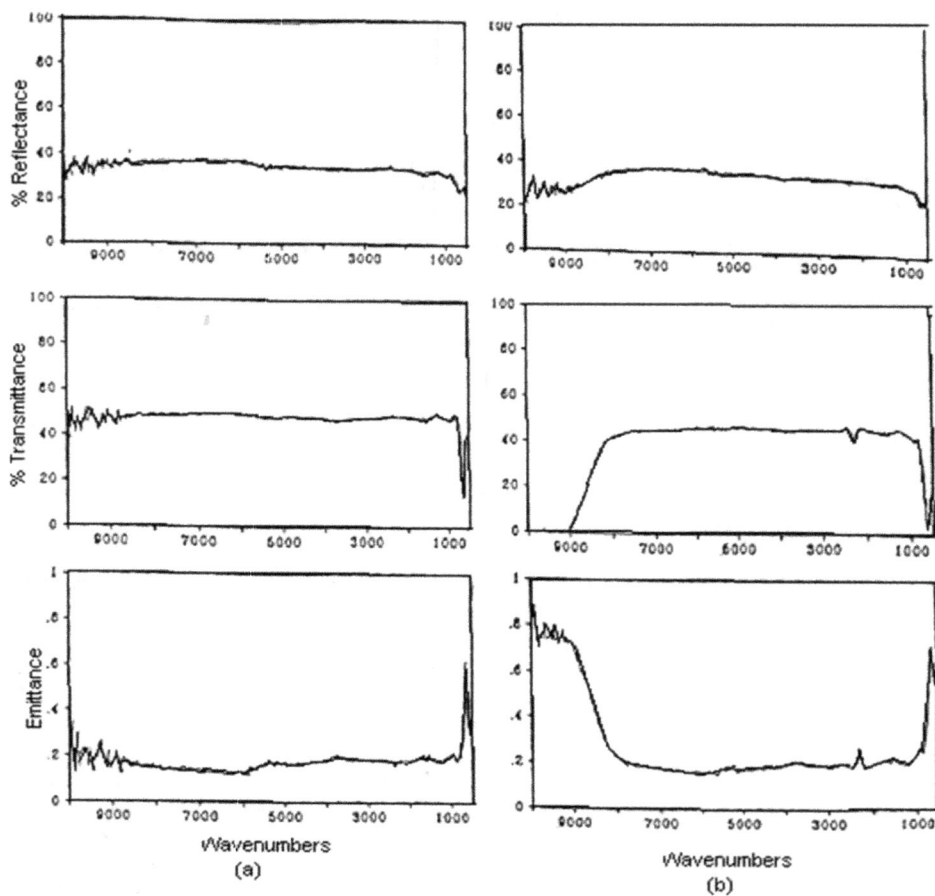

Figure 11.5. Temperature dependent optical properties of Fe-doped InP: (a) 40 °C and (b) 497 °C [1].

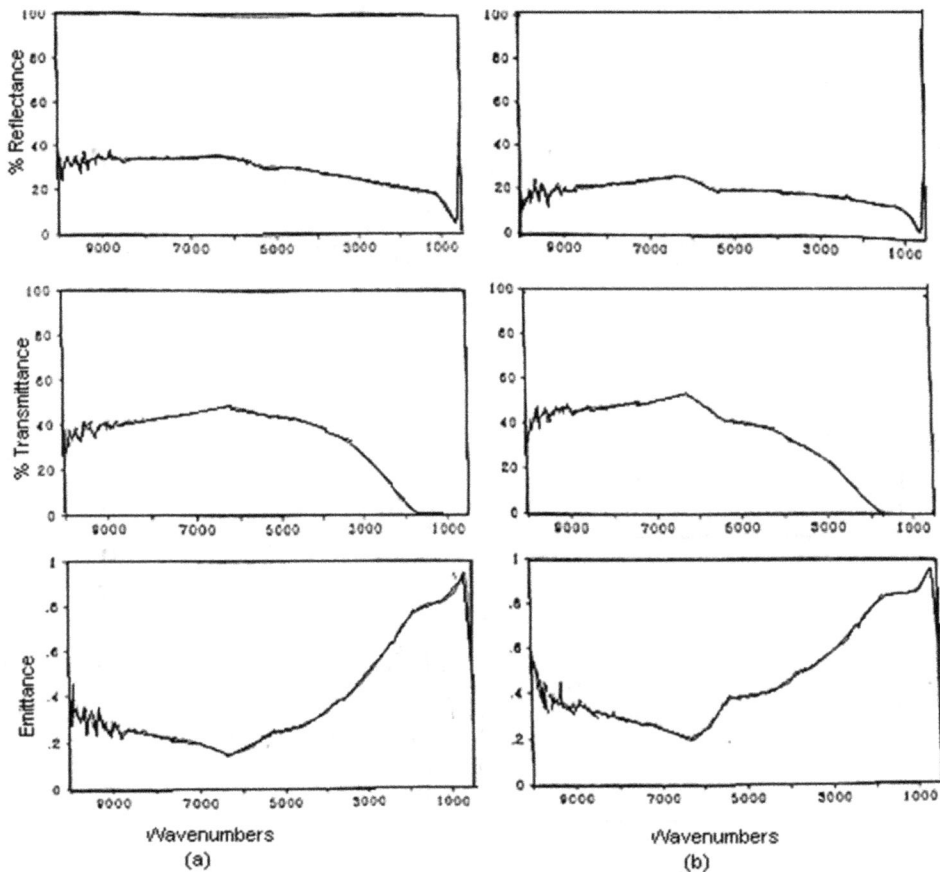

Figure 11.6. Temperature dependent optical properties of S-doped InP: (a) 99 °C and (b) 199 °C [1].

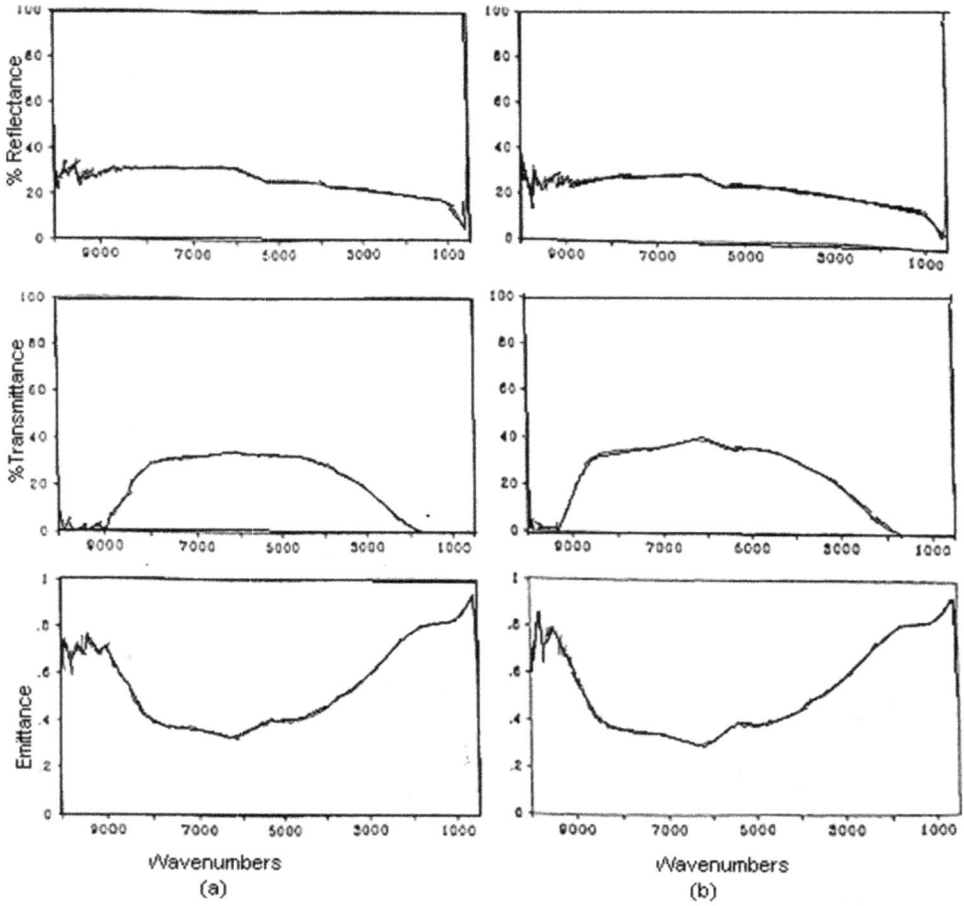

Figure 11.7. Temperature dependent optical properties of S-doped InP: (a) 350 °C and (b) 421 °C [1].

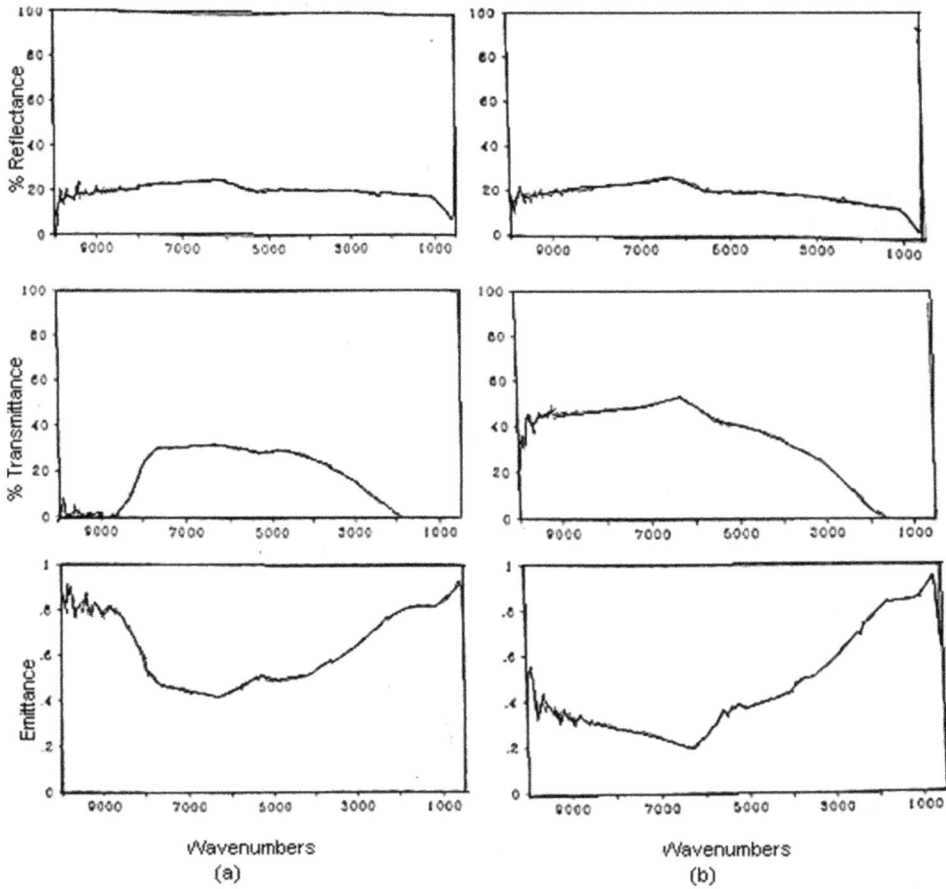

Figure 11.8. Temperature dependent optical properties of S-doped InP: (a) 519 °C and (b) 92 °C [1].

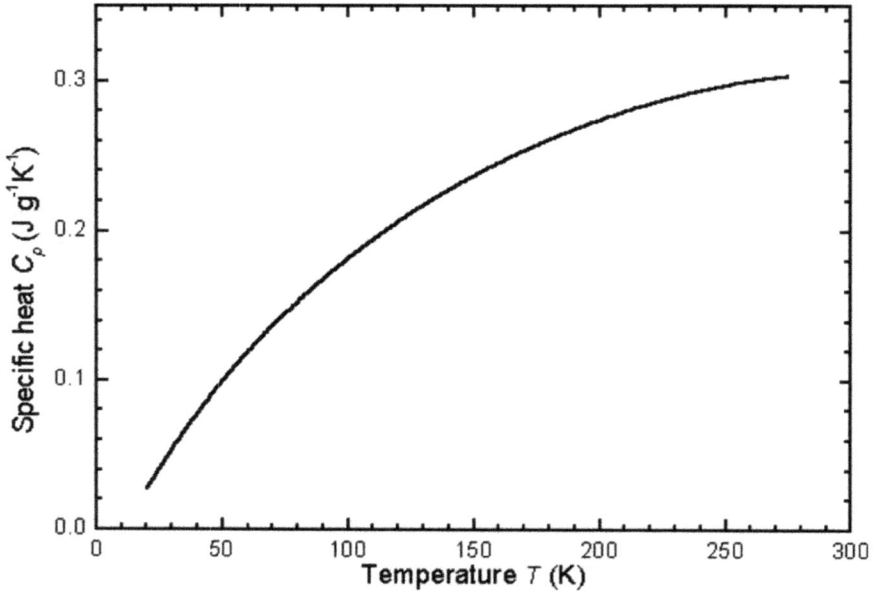

Figure 11.9. Temperature dependence of specific heat at constant pressure for InP [3].

The temperature dependence of specific heat at constant pressure for InP is shown in figure 11.9.

References

[1] Velagapudi R 1998 Investigation of optical properties of InP, AlN and sapphire for applications in non-contact semiconductor process monitoring *MS Thesis* NJIT
[2] http://www.ioffe.ru/SVA/NSM/Semicond/InP/thermal.html (accessed 10 April 2017)
[3] Piesbergen U 1963 Die durchschnittlichen Atomwärmen der AmBv-Halbleiter AlSb, GaAs, GaSb, InP, InAs, InSb und die Atomwärme des Elements Germanium zwischen 12 und 273 K *Z. Naturforschung* **18a** 141–7

Chapter 12

Cadmium telluride

Cadmium telluride (CdTe) is a II–VI compound semiconductor and an infrared optical window. CdTe is easier to manufacture when compared to silicon.

The linear thermal expansion coefficient (α) of CdTe is given by [1]:

$$\alpha = \frac{1}{l}\frac{dl}{dt}(4.9 \pm 0.1) \times 10^{-6} \ \mathrm{K^{-1}} \tag{12.1}$$

The thermal expansion coefficient of CdTe increases with increase in temperature. Below room temperature, α decreases, becoming negative at 62 ± 2 K. The plot of linear expansion coefficient of CdTe with temperature is shown in figure 12.1.

As the temperature decreases below 60 K, the linear expansion coefficient further decreases below zero and becomes more negative. The low temperature thermal expansion coefficient is plotted in figure 12.2.

The thermal conductivity of CdTe at 300 K is 0.075 W cm^{-1} K^{-1}.

The heat capacity of CdTe is measured by the dropping method. In order to understand its behavior, the values of heat capacity obtained from the experiments are summarized in table 12.1 [2].

After analyzing different sets of values from the literature, the values of heat capacity are summarized in table 12.2. The equation governing the heat capacity of CdTe and its temperature dependence, in the range of temperatures from 298.15 to 1365 K, is given by [2]:

$$C_\mathrm{p} = 50.28 + 6.22 \times 10^{-3} \ T - 185342.8 \ T^{-2} \tag{12.2}$$

The plot of heat capacity, as a function of temperature, of CdTe is presented in figure 12.3.

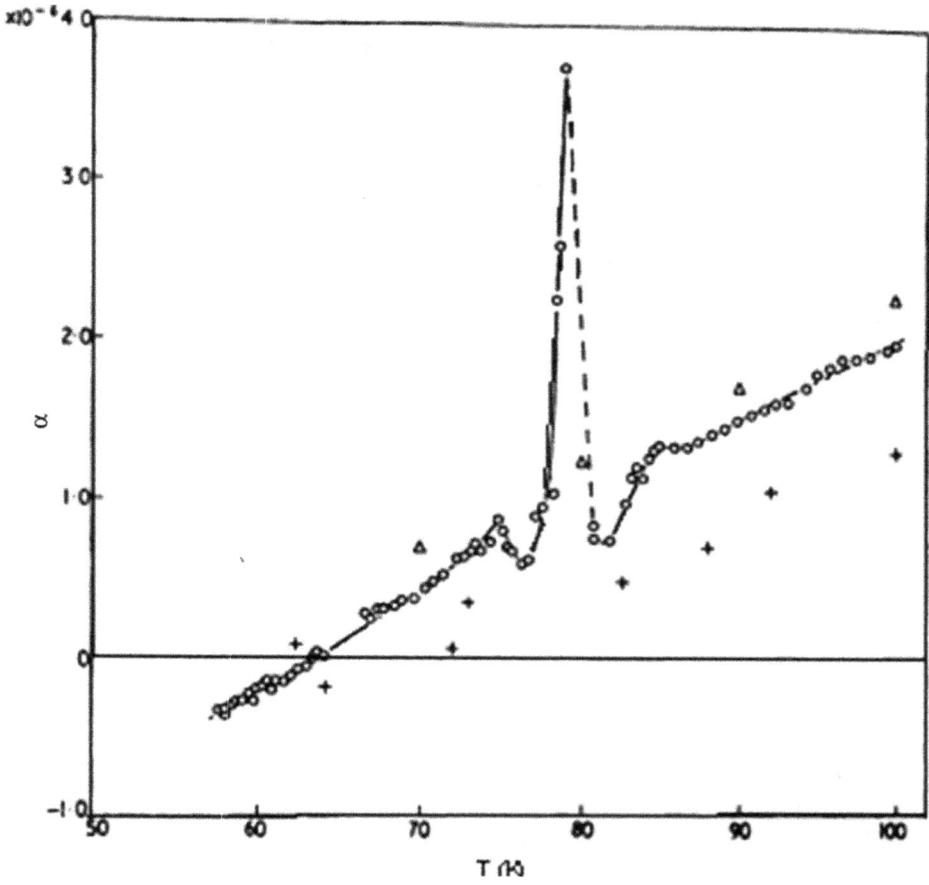

Figure 12.1. Linear thermal expansion coefficient α of CdTe versus temperature. Reproduced from [1] with permission of EDP Sciences.

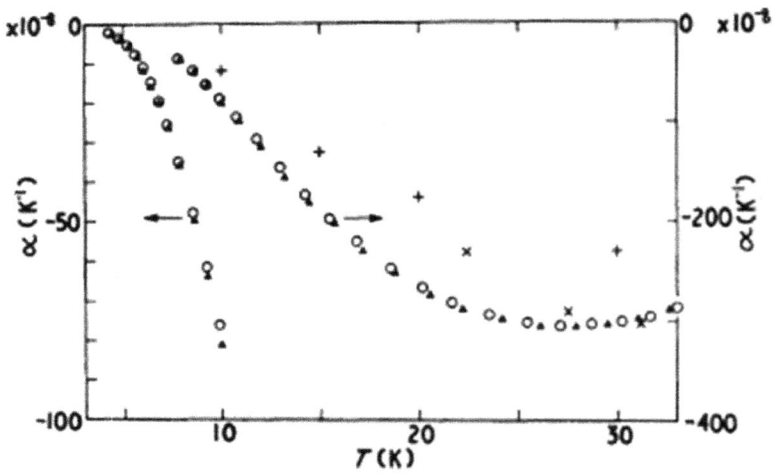

Figure 12.2. Linear expansion coefficient of CdTe versus temperature at lower temperatures.

12-2

Table 12.1. Values used to obtain specific heat of CdTe [2].

Method of statistical treatment	ΔT, K	Characteristic of the material	Error of measurement $H_T^0 - H_{298}^0$, %	C_p, J (mol K)$^{-1}$		Coefficients of equation $C_p^0 = a + bT - cT^2$		
				298 K	T_m, K	a	$b \times 10^3$	$c \times 10^{-5}$
MLS	473–965	contains traces of Sn	4.5	56.35	–	51.88	14.98	–
MLS	406–826	absent	5	49.88	–	39.64	21.72	–3.278
MLS	800–1373	impurity content of 10^{-3}% by mass	19–21	49.89	62.0	47.2	10.81	0.49
MLS	506–900	impurity content of less than 10^{-3}% by mass	4	51.18	–	51.18	–	–

The results of statistical treatment of the dependences $H_T^U - H_{298}^U = f(T)$ of solid CdTe, obtained experimentally by the 'dropping' method.

Table 12.2. Heat capacity of CdTe from 298 K to 1365 K [2].

T, K	C_p, J (mol K)$^{-1}$		
	Study 1	Study 2	Study 3
298.15	50.21	49.89	50.05
300	50.24	50.02	50.09
400	51.70	55.50	51.61
500	52.88	59.06	52.65
600	53.96	61.86	53.50
700	54.99	64.30	54.26
800	55.99	66.55	54.97
900	56.97	68.69	55.65
1000	57.95	70.77	56.32
1100	58.91	72.79	56.97
1200	59.88	74.79	57.62
1300	60.84	76.77	58.26
1365	61.46	78.04	58.67

Recommended values of heat capacity of cadmium telluride in the range from 298 K to the melting point according to the data of different studies.

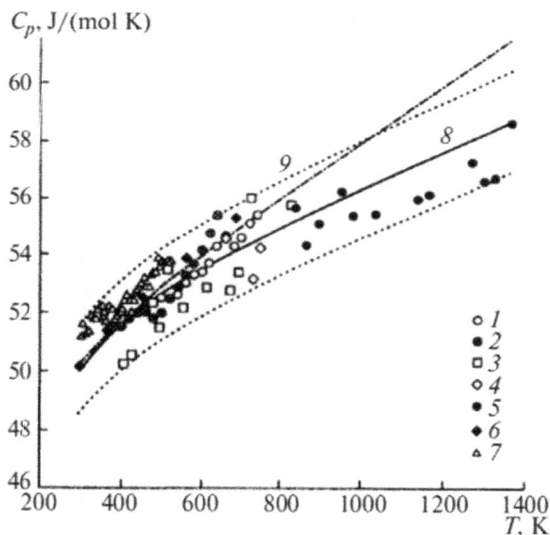

Figure 12.3. Heat capacity of CdTe versus temperature. Reproduced from [2] with permission of Springer.

Figure 12.4 shows the relative emissivity spectra for 0.85 mm thick CdTe wafer. The emissivity of CdTe sample is low at lower energies. The decrease in emissivity shifts to lower energy as the temperature increases. A slight increase in emissivity is observed at ~0.79 eV and 0.90 eV. This is attributed to defect related absorption in quartz window in the experimental apparatus.

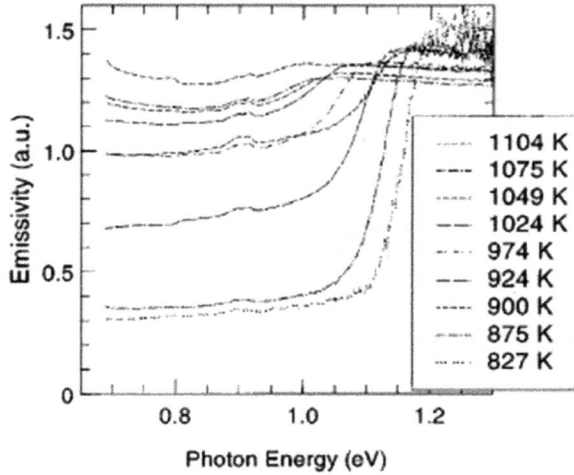

Figure 12.4. Relative emissivity spectra of a 0.85 mm thick CdTe wafer. Reprinted from [3] with the permission of AIP Publishing.

References

[1] Strauss A J 1977 The physical properties of cadmium telluride *Revue Phys. Appl.* **12** 167–84
[2] Pavlova L M, Pashinkin A S, Gaev D S and Pak A S 2006 The heat capacity of cadmium telluride at medium and high temperatures *High Temp* **44** 843–51
[3] Mullins J T, Carles J and Brinkman A W 1997 High temperature optical properties of cadmium telluride *J. Appl. Phys.* **81** 6374–9

Radiative Properties of Semiconductors

N M Ravindra, Sita Rajyalaxmi Marthi and Asahel Bañobre

Chapter 13

Mercury cadmium telluride

Mercury cadmium telluride (HgCdTe) is a pseudo binary alloy semiconductor of CdTe and HgTe, with zincblende structure. The composition of Hg and Cd can be chosen to tune the optical properties in the infrared region. HgCdTe is transparent in the infrared at photon energies below the energy gap. At present, HgCdTe is the most widely used variable band gap semiconductor for IR photodetectors.

Specific heat measurements of $Hg_{1-x}Cd_xTe$, taken from various sources in the literature, for different values of x, are shown in figure 13.1.

From figure 13.1, we can observe an anomaly in the form of abrupt increase in specific heat and departure from its normal linear behavior. This anomaly is attributed to the change in alloy composition of ternary alloys. A sharp increase in specific heat may be observed at temperatures beyond 690 K. Formation of point defects at higher temperatures contributes to this increase in specific heat. The specific heat of $Hg_{1-x}Cd_xTe$, below 300 K, increases with temperature and shows very little composition dependence, except for a possible alloy hardening effect that slightly depresses at values near 220 K.

The thermal conductivity (k) of $Hg_{1-x}Cd_xTe$, like any other semiconductor, depends on the purity of the sample. The temperature dependence of the thermal conductivity is given by the relation [1]:

$$k = 100\frac{A}{T^n} \tag{13.1}$$

where k is thermal conductivity, A and n are factors depending on x, and T is temperature in Kelvin.

doi:10.1088/978-1-6817-4112-3ch13

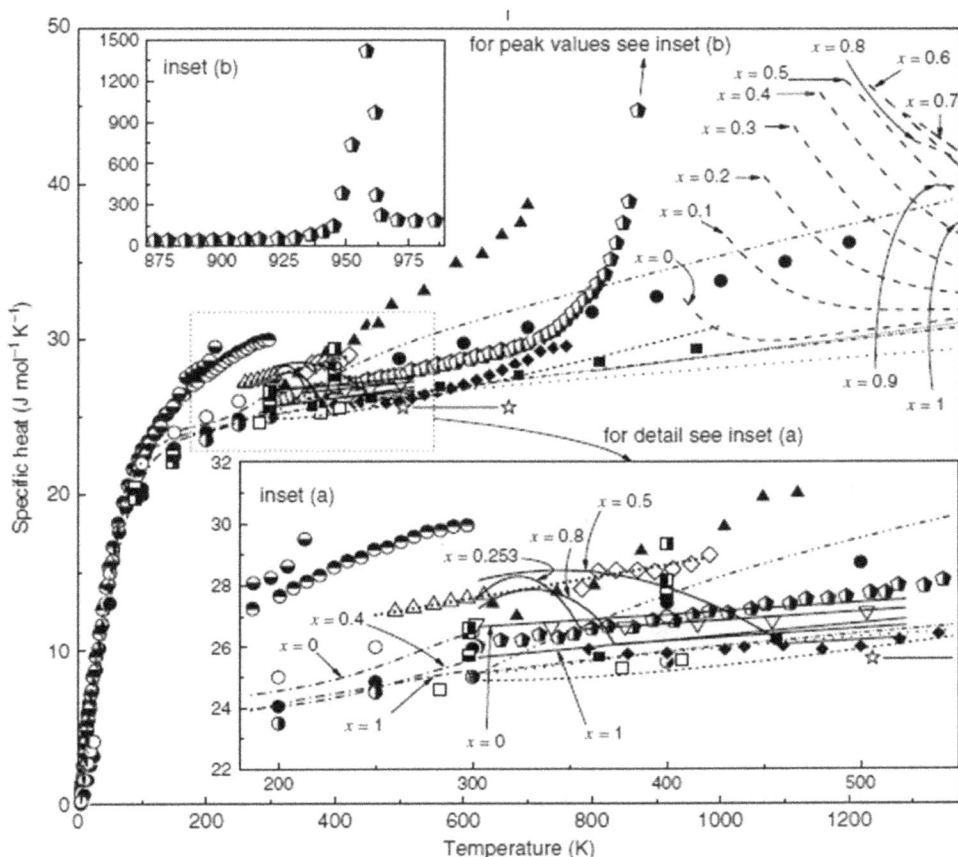

Figure 13.1. Constant-pressure specific heat, C_p, of $Hg_{1-x}Cd_xTe$ as a function of temperature as measured by different sources. Reprinted from [1] with the permission of AIP Publishing.

Temperature dependence of the thermal conductivity of $Hg_{1-x}Cd_xTe$ is shown in figure 13.2.

Measurements of the temperature dependence of the thermal conductivity indicate a maximum near 8 K, common to all compositions, with the thermal conductivity generally indicating a decreasing trend with increasing temperature up to the solidus.

Thermal conductivity is also dependent on the purity of the sample just like specific heat, even at cryogenic temperatures.

Far infrared reflection spectra of HgCdTe are shown in figure 13.3. The characteristic features in reflection bands for the alloys with $x = 0.45$ and $x = 0.18$ are HgTe-like and CdTe-like as well. For large values of x, fine structure in CdTe-like reflection band may be observed. These features become more pronounced at lower temperatures. Another significant feature in the reflection band

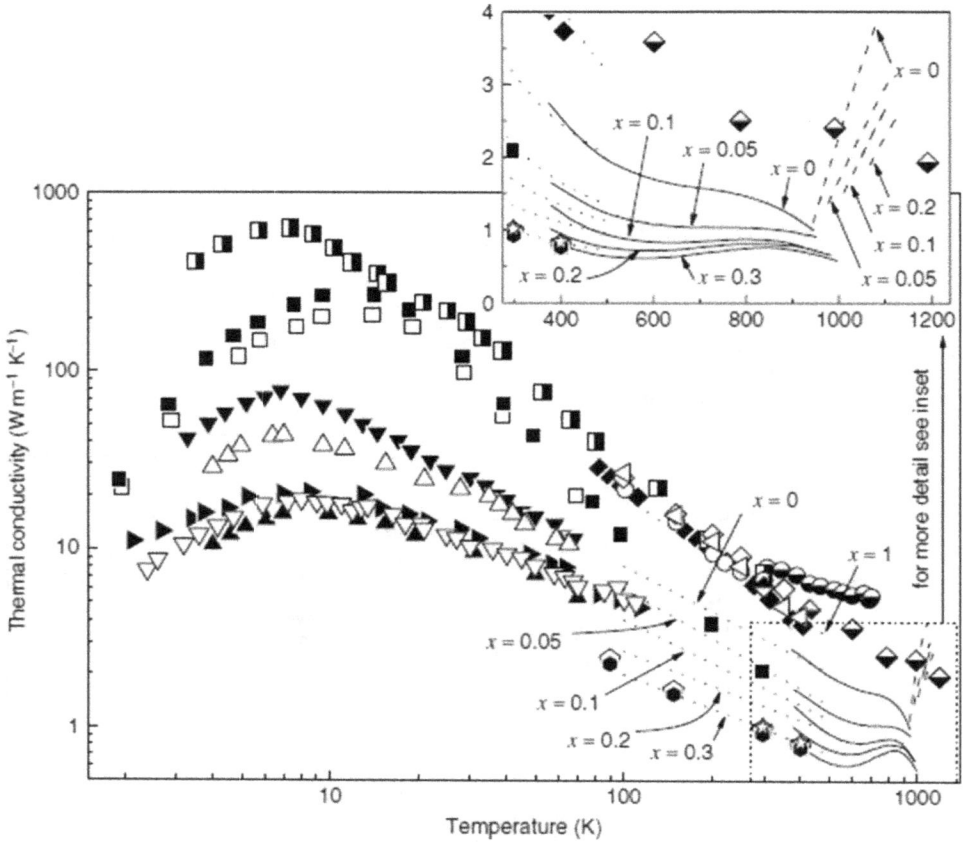

Figure 13.2. Thermal conductivity of $Hg_{1-x}Cd_xTe$ as a function of temperature. Reproduced from [1].

near 130 cm^{-1} ($\sim$77 μm) for the sample with small x value is the presence of a complicated structure at room temperature and higher temperature [1].

Figure 13.4 shows the emittance of HgCdTe for two compositions (LWIR—long wavelength infrared and SWIR—short wavelength infrared). The emissivity of HgCdTe is $\sim$0.7 and is constant, for both the samples, in the wavelength region of 1.5 to 3.5 μm at room temperature. The constant emissivity is attributed to the intrinsic nature of HgCdTe. At room temperature, the emissivity of HgCdTe is 0.68 [2].

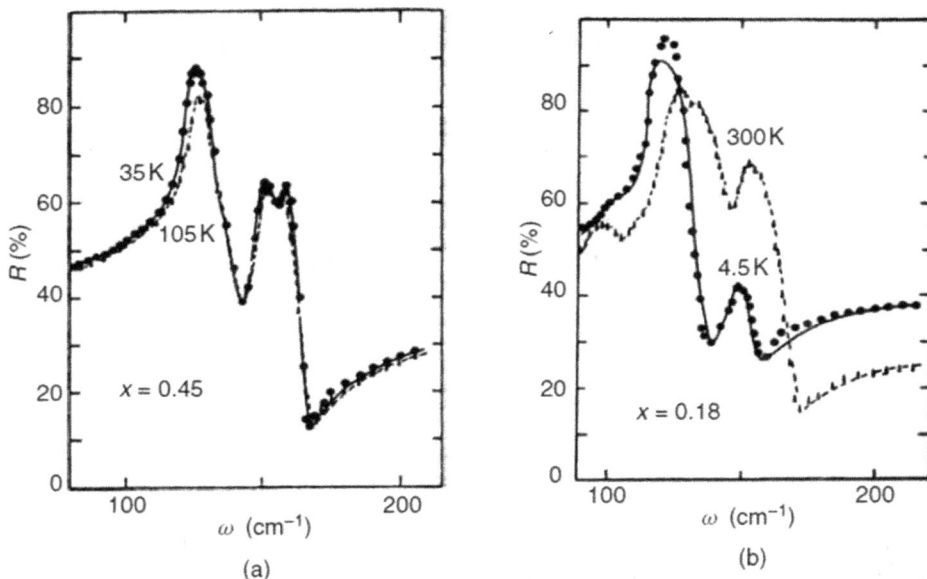

Figure 13.3. Far-infrared reflection spectra for $Hg_{1-x}Cd_xTe$ sample. (a) $x = 0.45$ at 35 K (dots: experiment; solid line: a fit) and 105 K (triangles: experiment; dashed line: a fit). (b) $x = 0.18$ at 4.5 K (dots: experiment; solid line: a fit) and 300 K (triangles: experiment; dashed line: a fit). Reproduced from [1].

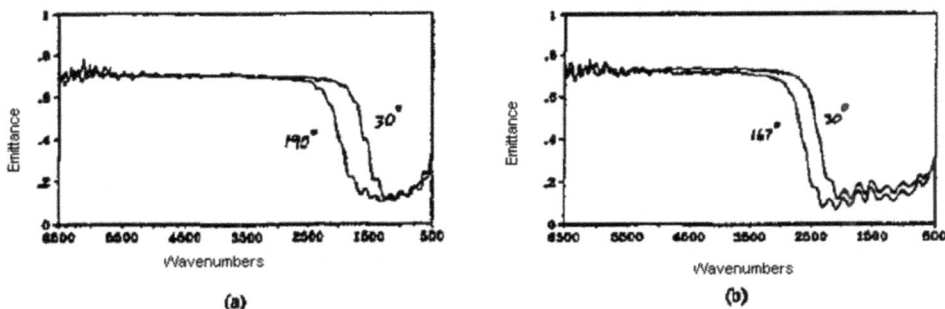

Figure 13.4. Spectral emittance of HgCdTe of two compositions (LWIR and MWIR) (a) LWIR at temperatures 30 °C and 190 °C, (b) MWIR at temperatures 30 °C and 167 °C [2].

References

[1] Capper P and Garland J W 2010 *Mercury Cadmium Telluride: Growth Properties and Applications* (Chichester: Wiley)

[2] Ravindra N M, Tong F M, Kosonocky W F, Markham J R, Liu S and Kinsella K 1994 Temperature dependent emissivity measurements of Si, SiO_2/Si and HgCdTe *Materials Research Society Symp. Proc.* **342** 431–6

Chapter 14

Modeling

The rapid advancements in semiconductors have warranted introduction of various complex integrated circuits with complicated structures for a variety of applications. This has created an increased demand to study and optimize the properties of these structures at the micron and sub-micron level. Introduction of modeling in designing and optimizing the device structures and the material properties helps to better understand their integration and reduce the cost of manufacturing.

In the literature, several models have been successfully used to simulate the radiative properties of semiconductors. Some of the models have been described here.

14.1 MultiRad

MultiRad is a PC-based software designed by Hebb *et al* [1] at the Massachusetts Institute of Technology. MultiRad facilitates the calculation of the radiative properties of multilayer stacks. The calculation of the optical properties is based on thin film optics. This model assumes that the layers are optically smooth and parallel, and also that the materials are optically isotropic. The inputs are wavelength of incident light, angle of incidence, thickness of each layer and optical constants, corresponding to the wavelength, of each layer. The output is spectral reflectance, transmittance and absorptance—often equal to emittance. These properties are integrated with respect to wavelength to obtain spectral properties as well as the total optical properties.

MultiRad implements the matrix method of multilayers to calculate the radiative properties of thin films. Figure 14.1 shows the sample screenshot from MultiRad.

14.2 RadPro

Georgia Institute of Technology and the National Institute of Standards and Technology (NIST) have developed a software package to predict the radiative properties of silicon wafers in rapid thermal processing (RTP) environments [2].

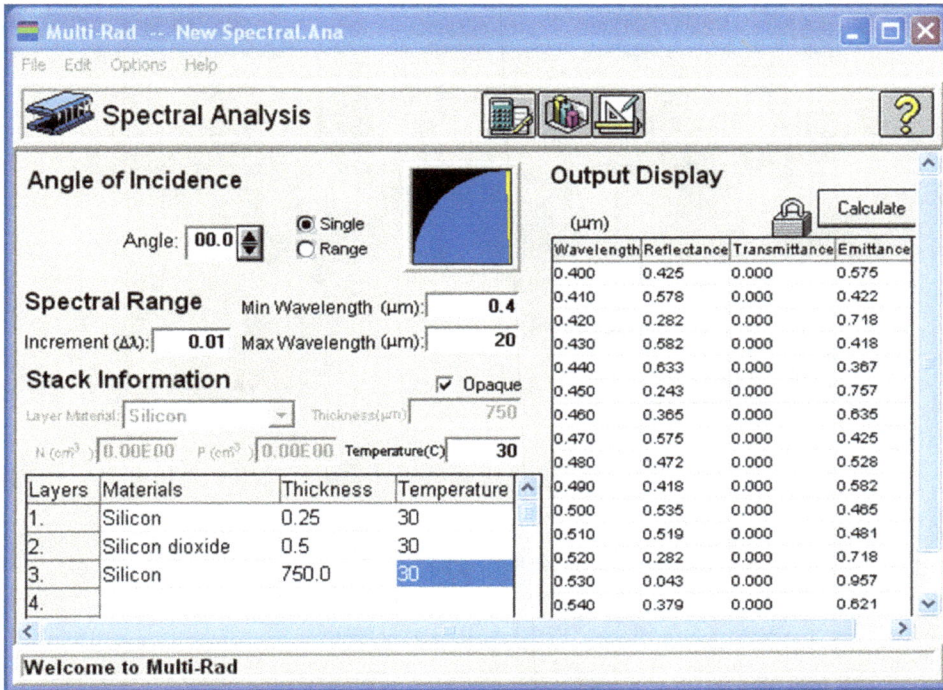

Figure 14.1. Spectral analysis window of MultiRad. Reproduced from [1] with permission of J Hebb.

This software allows the simulation of the directional, spectral and temperature dependence of the radiative properties for multilayer structures of silicon and its related materials such as silicon nitride, silicon dioxide and polysilicon. User defined materials can be added easily by inputting their respective wavelength dependent optical constants.

RadPro provides an option to use coherent, incoherent and opaque materials without spectral integration. The allowed input options include wavelength, temperature, or angle of incidence. RadPro provides two options—opaque substrate and hemispherical emittance options in order to maximize its applicability. Figure 14.2 shows an example of the calculation of the radiative properties of a silicon wafer.

14.3 PV optics

PV optics is a software package that permits the modeling of the optical properties of solar cells [3]. The software also simulates the efficiency of the selected device configuration and helps to improve the device design. PV optics facilitates the analyses of multilayer structures for solar cell applications. The layers may be semiconductors, dielectrics, metals, etc. The criterion to select the optics viz ray optics or wave optics, is the thickness of the layers in comparison to the wavelength of incident light. The output includes plots of reflectance, transmittance and absorptance spectra. PV Optics can also help gain insight into the mechanism of

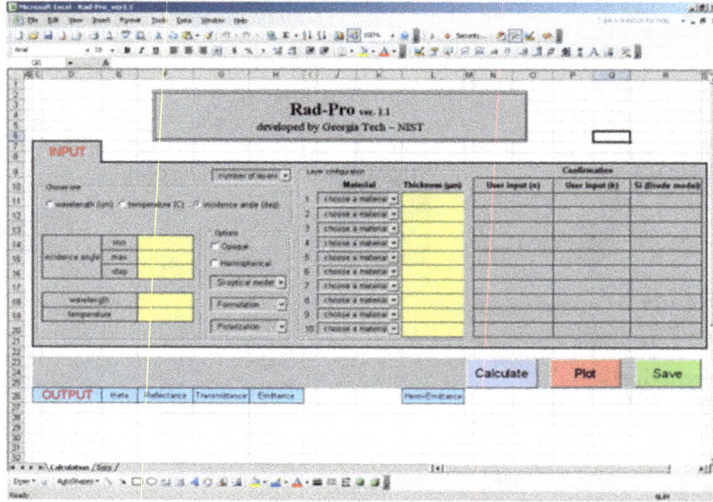

Figure 14.2. Example of calculation of the radiative properties of a silicon wafer with a silicon dioxide film. Copyright 2005 IEEE. Reprinted, with permission, from [2].

Selection of Device Configuration

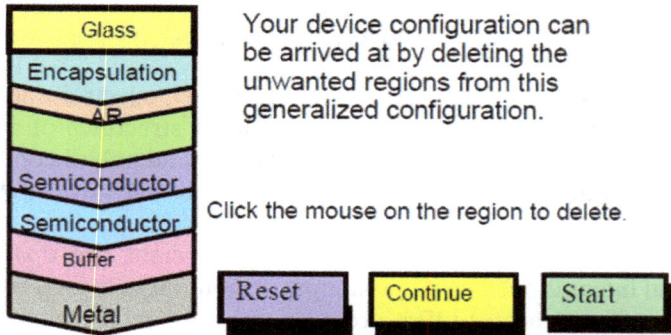

Figure 14.3. Illustration of PV optics 'Selection of Device Configuration' [3].

losses in photovoltaic devices, making it a useful design for solar cell applications. Figure 14.3 shows the illustration of PV optics.

14.4 Ray tracing

Ray tracing is an extension of PV optics with enhanced capabilities [4]. In addition to the features included in PV optics, ray tracing permits the calculation of emissivity (equivalent to the absorptivity). The modifications in ray tracing are as follows:

- Extension of wavelength range to include applications in pyrometry and RTP
- Calculations of refractive index and extinction coefficient of Si at various temperatures.

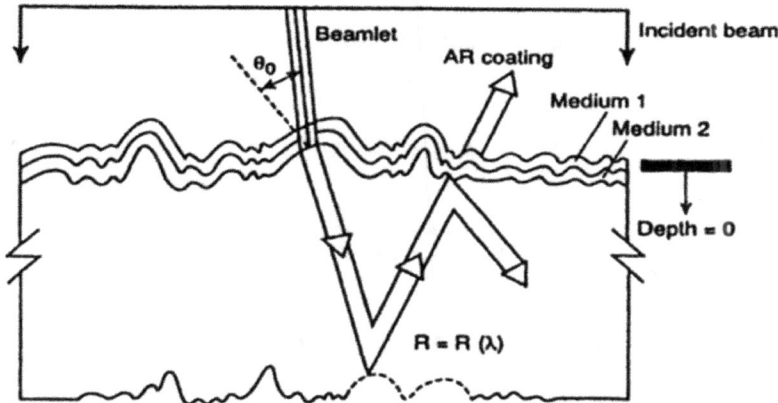

Figure 14.4. Illustration of the methodology of ray tracing calculations. Reproduced from [4] with permission of Springer.

Figure 14.4 shows an illustration of the methodology of ray tracing calculations. The model calculates total absorption, reflection, transmission and scattering effects due to nonplanar surfaces.

References

[1] Hebb J P 1997 Pattern effects of rapid thermal processing *PhD Thesis* Massachusetts Institute of Technology, Cambridge, MA
[2] Lee B J and Zhang Z M 2005 Rad Pro: Effective software for modeling radiative properties in rapid thermal processing *13th Int. Conf. on Advanced Thermal Processing of Semiconductors (RTP 2005)* pp 275–81
[3] Sopori B, Madjdpour J, Zhang Y and Chen W 1999 PV Optics: an optical modeling tool for solar cell and module design *Electrochemical Society Int. Symp. (Seattle, Washington, 2–6 May)*
[4] Ravindra N M, Ravindra K, Sundaresh Mahendra S, Sopori B and Fiory A T 2003 Modeling and simulation of emissivity of silicon-related materials and structures *J. Electron. Mater.* **32** 1052–8

Chapter 15

Applications

By definition, a semiconductor has electrical conductivity in between a conductor and an insulator. There are several types of semiconductors which are classified by their electronic properties that determine their use and applications. Semiconductors are elements such as Si and Ge or compound semiconductors such as SiC, GaAs, GaN, InP, InSb, CdTe and HgCdTe [1].

Semiconductor devices are critical to the consumer electronics industry. The products include mobile phones, computers and audio-visual systems among others. Semiconductor devices have replaced thermionic devices (vacuum tubes), by exploiting the electronic properties of semiconductors, principally silicon, germanium and gallium arsenide. Semiconductor devices are manufactured as single discrete devices and as integrated circuits (ICs).

Each semiconductor material has advantages and disadvantages. The mechanical, electronic and optical properties of the semiconductor determine their applications. Optical semiconductor devices include photocells, solar cells, phototransistors, light activated switches, optical transmission windows, light-emitting diodes (LEDs) and infrared LEDs that are developed for applications in wireless transmission and reception. Due to their low luminous efficiency, silicon and germanium are mainly used for photodetectors, while compound semiconductors, with their high luminous efficiency, are used for LEDs and laser diodes.

The interaction of optical radiation with matter results in several light induced processes such as photovoltaic, photoconductive, Dember, photon drag and photo-electromagnetic effects. Based on these photo-effects, a wide variety of optical devices have been developed with the use of semiconductors depending on their optical properties.

15.1 Silicon

Silicon (Si) has been the most widely used material in the semiconductor industry for over 50 years. The low cost of the material, in combination with its optical and

doi:10.1088/978-1-6817-4112-3ch15

electrical properties and its easy integration with planar microfabrication technology, makes silicon the best candidate among the various semiconductor materials for a variety of applications. Typical optoelectronic applications of silicon include lenses, optical filters, waveguides, thermal imagers and solar cells.

Silicon is not an efficient material for light detection because it does not absorb radiation above 1 μm, even though some approaches such as two-photon absorption, internal photoemission effect and sub-band gap photo-effect have been proposed; the generation and recombination of electron–hole pairs in silicon leads to a high noise level [2]. The indirect band gap of silicon ~1.16 eV makes it unsuitable to be used as a light emitter [3].

Silicon photodiodes are the best understood optical devices, covering the range below 1 μm, that covers near infrared, visible light, x-ray and gamma rays [4, 5]. The principal types of silicon photodiodes are p–n junctions, p–i–n junctions, UV-blue enhanced photodiodes and avalanche photodiodes [6].

Silicon p–n junction photodiodes operate as photovoltaic or photoconductive, depending on the operation mode—unbiased or reverse biased, respectively. The fabrication and development of photovoltaic cells by the solar industry is based on silicon (p–n junction). Silicon solar cells are the most suitable solution for mass production of photovoltaic (PV) modules [7]. The first silicon solar cell was developed in the 1950s by Bell Labs with a conversion efficiency of around 4.5%. In the last six decades, the improvement of the cell design has matured to an improvement in efficiency to close to 20%. In actuality, solar cells are fabricated using crystalline, amorphous and polycrystalline silicon; these solar cells have different efficiencies and fabrication costs that determine their consideration in the fabrication of the photovoltaic (PV) modules [8]. A typical PV module is shown in figure 15.1, with complete interconnected solar cells [9].

Figure 15.1. Photovoltaic module with complete interconnected solar cells [9].

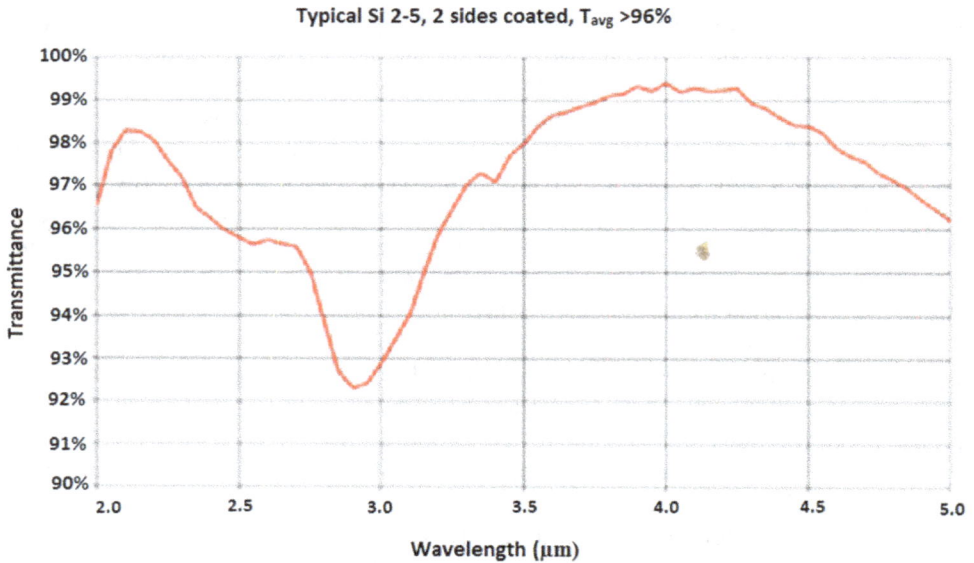

Figure 15.2. Silicon infrared transmittance window band [11].

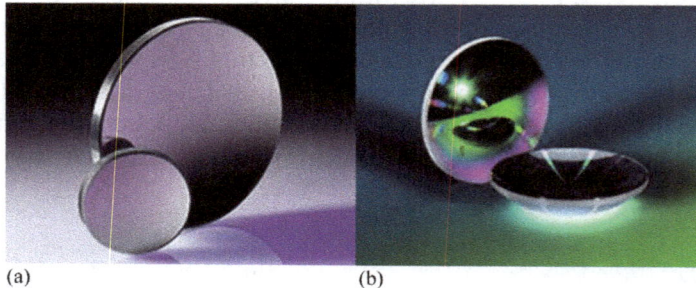

Figure 15.3. Silicon lenses: (a) plano-convex (PCX) and (b) aspheric [12, 13].

Silicon is widely used in the fabrication of optical lenses for a variety of applications such as focusing or diverging optical radiation and transmission from optical windows. Silicon is ideal for transmission in the infrared spectrum. With an index of refraction ($n > 3$) [10], silicon provides a good transmittance in the 2–5 μm spectral band. This facilitates in the fabrication of silicon based infrared filters in optical systems that require high infrared transmittance. Figure 15.2 shows the silicon infrared optical transmission window band [11].

Figure 15.3 shows two examples of silicon lenses manufactured by Edmund Optics Inc. Silicon plano-convex (PCX) lenses are ideal for near-infrared (NIR) (1.2–7 μm) imaging and infrared spectroscopy, due to their low cost and density. These lenses are offered with a variety of focal lengths from f/1 to f/10. Silicon aspheric lenses are designed to correct spherical aberrations and are designed to withstand high levels of temperature and pressure, as a consequence of the excellent mechanical and thermal properties of silicon [12, 13].

15.2 Germanium

Historically, germanium (Ge) has been used in the fabrication of early semi-conductor devices such as radar detection diodes and the first transistors. Its high thermal sensitivity is a disadvantage in comparison with the more stable thermal properties of silicon. Germanium has a broader mid-infrared transparency range and higher carrier mobility and larger non-linear effects when compared to Si [14].

In germanium, the two most used photodiodes are p–n junction and p–i–n junction. Because of its narrow band gap (0.66 eV), the UV-blue range is not applicable to germanium detectors. Like silicon, germanium is widely used in the fabrication of optical components such as lenses, infrared filters and thermal imagers.

High-purity germanium (HPGe) counting detectors have excellent energy reso-lution. They are used in many detection fields, from nuclear and astro-particle physics to homeland security and environmental protection [15, 16]. Germanium detectors have been developed for particle tracking, x-ray position sensing and spectroscopy, based on the integration with a read-out system and a cryostat [17].

Figure 15.4 shows a photograph of a chip (infrared detector array), an enlarged portion with a few pixels and the electronics of an individual pixel and its schematic cross section. The detectors are fabricated by depositing Ge on the Si n-well, after opening a suitable via. Anode and cathode are obtained by the n-well contact and a p$^+$ diffusion (the parallel Si diode being unimportant due to its lack of sensitivity to near infrared wavelengths) [18].

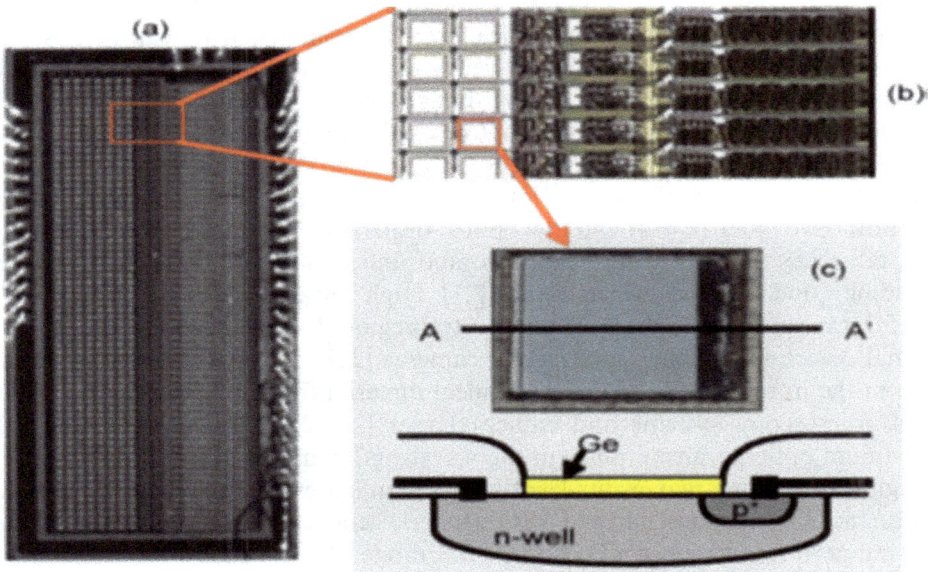

Figure 15.4. (a) Photograph of the whole chip. (b) Enlargement showing a few pixels and the row of electronics. (c) Photograph of a pixel and sketch of its cross section [18].

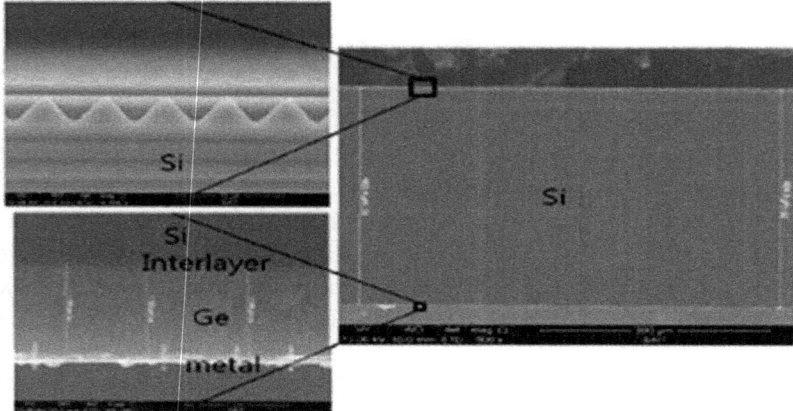

Figure 15.5. TEM image of fabricated stack junction (right), inverted pyramid texturing of Si (top left), and Ge cell(bottom left) [20].

Germanium is one of the most important semiconductors used today in the fabrication of high efficiency thermophotovoltaic (TPV) and photovoltaic (PV) [19] solar cells. A single junction of silicon solar cell has a practical limited efficiency of 26%. One of the solutions to overcome this limiting efficiency is the use of stack junctions with the help of germanium. Germanium, with a band gap of 0.66 eV, absorbs the photons that cannot be absorbed by the Si layer. Ge is a single element material easy to purify and is easy to dope with n-type or p-type impurities. An example of the device is the Si/Ge stack configuration, a 21.3% Si/1.6% bottom Ge stack junction with 22.9% module efficiency. The solar spectrum can be absorbed more effectively by combining the efficiency of multi-junctions. Figure 15.5 shows the transmission electron microscopy image of a double junction Si/Ge, the interdigitated front contact (IFC) on top of the Si layer and the back contact (BC) at the bottom of the Ge layer [20].

Germanium is an excellent material for the fabrication of immersed lens due to its high refractive index ($n = 4$) and low optical dispersion [10]. Germanium is used for high efficiency anti-reflection coatings and infrared filters in optical systems providing good infrared transmittance [21]. High refractive index of Ge provides a useful transmittance in the 2–23 μm spectral region. It is used as an optical filter in thermal detectors and thermal imaging cameras [22]. The transparency of germanium in the infrared band makes it an ideal material for the fabrication of optical fibers, microscopy systems and camera lenses [23, 24]. Figure 15.6 shows the germanium optical transmittance in the wavelength window of 3–12 μm [25].

Figure 15.7 shows two examples of germanium lenses manufactured by Edmund Optics Inc. Germanium infrared (IR) aspheric lenses provide diffraction limited focusing performance over a broad spectral range of mid- and long-wave infrared spectra [26]. Germanium infrared achromatic lenses are ideal components for designers and researchers working in the 3–5 μm spectral region [27].

Typical Ge 3-12, 2 sides coated, T_{avg} > 94%

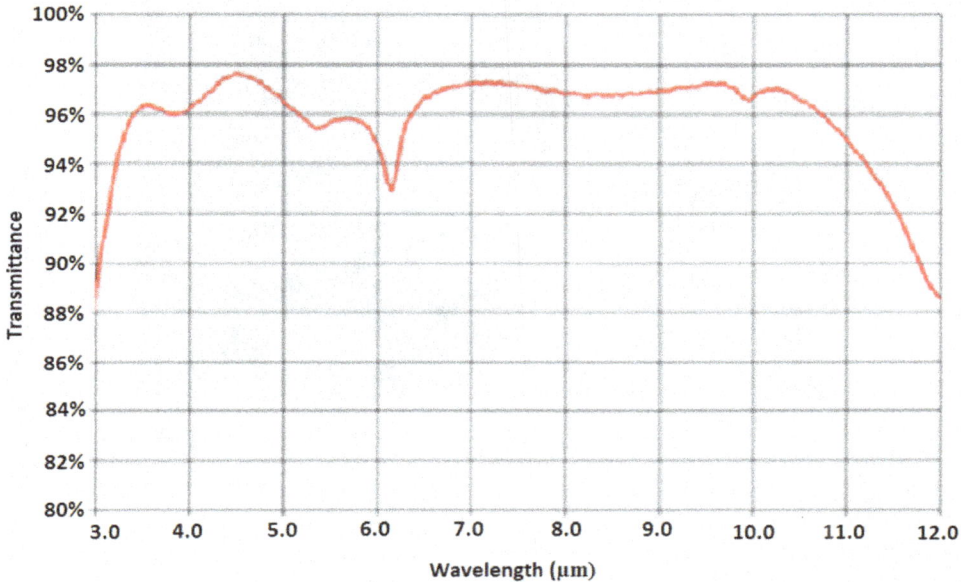

Figure 15.6. Germanium infrared transmittance window band [25].

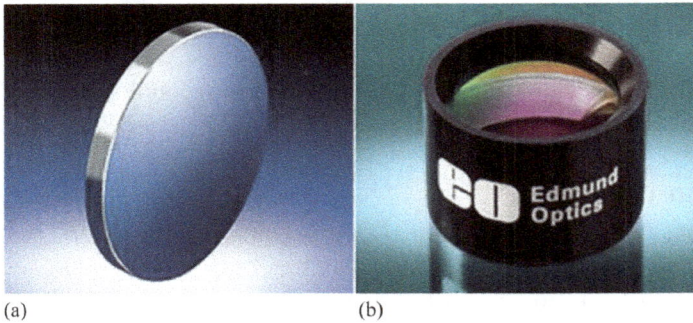

(a)   (b)

Figure 15.7. Germanium lenses: (a) aspheric lens and (b) achromatic lens [26, 27].

15.3 Graphene

Graphene is a transparent and flexible two-dimensional conductor with unique optical properties that holds great promise for various applications in products such as lenses, solar cells, flexible light-emitting diodes (LEDs), liquid crystal displays (LCDs), touch screens and smart windows for smartphones [28–31]. The ultrathin coating of graphene monolayer absorbs 97% of the incident light across a broad-band spectrum, from the UV to the infrared. It is anticipated that the optical properties of graphene will revolutionize the display industry in the manufacture of smartphones, display screens and wearables.

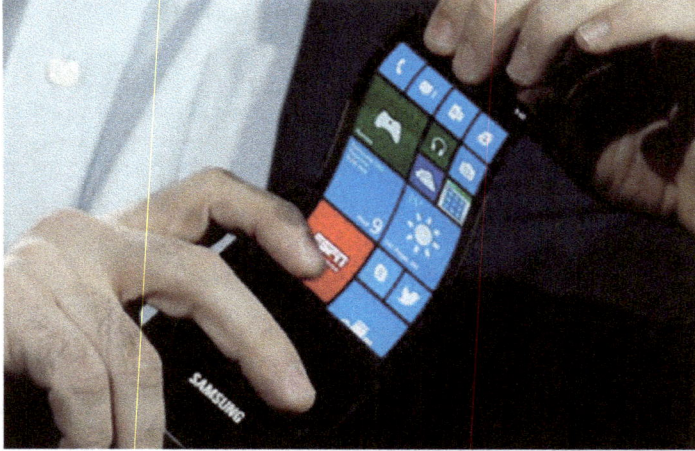

Figure 15.8. Bendable and flexible graphene touch-screen display (Samsung Youm) [33].

During the last decade, leading companies in the mobile equipment and television manufacturing sector such as Apple, LG Electronics, Nokia, Panasonic, Samsung Electronics and Sony have been working on the development of flexible organic light emitting diodes (OLEDs)-based touch-screens. China-based 2D-Carbon Graphene Material Co. Ltd manufactures commercial graphene-based touch panel modules for applications in the fabrication of smartphones, wearable devices and home appliances [32].

In the 2013 annual trade show, Consumer Electronics Show (CES), organized by the Consumer Technology Association, Samsung Electronics introduced their concept of a bendable flexible phone called 'Youm'. This new technology allows phones and TVs to be thin and flexible with a high quality image [33], as can be seen in figure 15.8.

15.4 Silicon carbide

Silicon carbide (SiC) provides significant advantages for the manufacture of fast, high-temperature and high-voltage devices [34]. Electronic applications of silicon carbide such as light-emitting diodes (LEDs) and detectors, in early radios, were first demonstrated around 1907.

The phenomenon of electroluminescence in SiC was discovered in 1907 [35]. The first commercial LEDs, based on SiC, were manufactured in the Soviet Union in the 1970s, and were made from 3C-SiC. Later, in the 1980s, blue LEDs, based on 6H-SiC, were produced worldwide. Today, SiC is one of the important materials that is used as a substrate for the growth of GaN devices and for efficient thermal management in high-power LEDs [36].

Cree Inc., a US-based company, is the largest manufacturer of silicon carbide substrates, lighting-class LEDs, lighting products, products for power and radio frequency (RF) applications. Cree has developed a SiC based SC^3 Technology™ platform. It has been demonstrated to yield high performance products for applications that require high endurance for semiconductor devices operating at

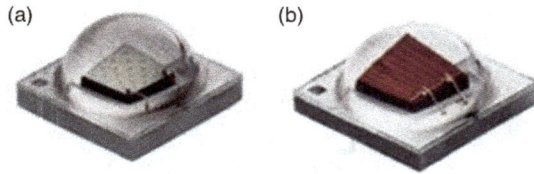

Figure 15.9. Cree high power XLamp® XP-E2 LEDs based SiC, a) single color blue, 485 nm max and b) single color red, 73.9 lm [38].

high temperatures or high voltages [37]. Figure 15.9 shows examples of Cree XLamp® XP-E2 LEDs based on SiC, with good performance and delivery of high lumens per watt. XP-E2 LEDs exhibit high emissivity efficiency and are suitable for the manufacture of LEDs retrofit lamps, outdoor, portable, indoor, directional, emergency vehicle or architectural applications [38].

Silicon carbide has low thermal expansion coefficient, high hardness, rigidity and thermal conductivity, making it a desirable mirror material in the fabrication of astronomical telescopes. With the mission to study the Universe, the European Space Agency (ESA), has constructed the Herschel Space Observatory and the Gaia space observatory spacecraft [39]. The Herschel Space Telescope is equipped with SiC mirrors [40]. Also, Gaia space observatory spacecraft subsystems are mounted on a rigid silicon carbide frame, which provides a stable system structure that will not contract or expand due to changes in temperature.

The Herschel Space Observatory, with a primary mirror diameter of 3.5 m, is the largest single mirror ever built for use in space and the largest infrared space observatory launched to date [41]. Gaia is a space observatory designed for astrometry; its mission is to chart a high definition/resolution/precision 3D space map [42]. Both of these projects have been developed to study the far infrared to sub-millimeter parts of the spectrum (from 30 to 1000 μm) in the Universe. They will help to explore parts of the infrared spectrum that are difficult to detect from the Earth's surface as a result of the atmospheric absorption bands.

15.5 Gallium arsenide

Gallium arsenide (GaAs) is a III–V compound semiconductor with a direct band gap of 1.42 eV at 300 K. The electron and hole mobilities in GaAs are higher than that of silicon. Its high breakdown voltage and excellent optical properties position GaAs as a material with numerous applications; these include the manufacture of infrared light-emitting diodes (LEDs), solar cells (photovoltaic), thermal imagers, laser diodes and infrared transmission optical windows.

Gallium arsenide is considered as an alternative material to silicon because of its electronic and optical properties. The cost of manufacturing process as well as the control of defects and wafer scaling is still a disadvantage for GaAs. Gallium arsenide single-junction and multi-junction solar cells can achieve high efficiency levels in the range of 20%–30%, making it an excellent candidate to be part of a multilayer stacked structure that collects radiation over the entire visible spectrum

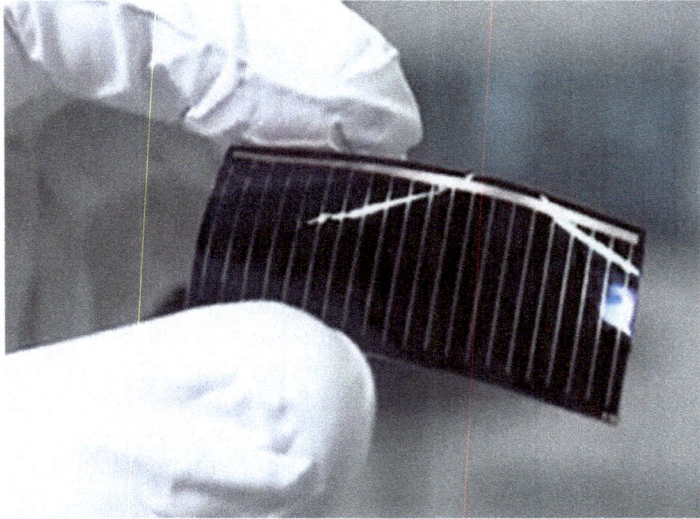

Figure 15.10. Flexible gallium arsenide thin film solar cell (Alta Devices) [44].

range [43]. Alta Devices has developed single-junction GaAs solar cells with high performance power efficiencies of 28.8% for single-junction and 31.6% for dual-junction cells [44]. Figure 15.10 shows a flexible GaAs thin film solar cell manufactured by Alta Devices.

Typical applications of gallium arsenide include lenses, prisms and infrared transmission windows. Gallium arsenide, with an index of refraction (n) >3, is an infrared window in the 2–15 μm spectral band [45, 46]. Figure 15.11 shows the gallium arsenide infrared transmission band (0.9–20 μm) [46].

Infrared LEDs, having a high output power that is invisible to humans, are used as radiation source for optical emitters and sensors. GaAs is used to manufacture devices for applications in infrared communication and infrared imaging systems. In combination with other elements, ternary compounds of GaAs are used in the manufacture of visible LEDs that emit radiation in the visible spectrum. These include the following: AlGaAs 610 < λ < 760 μm (red), GaAsP 570 < λ < 610 μm (orange and yellow). Figure 15.12 shows two examples of commercially available LEDs manufactured by Hamamatsu Photonics—a model TO-46 style GaAs infra-red LED of high reliability and radiant output power [47], and a high power red emitting diode (AlGaInP) [48].

15.6 Gallium nitride

Gallium nitride (GaN) is a wide bangap compound semiconductor, with a direct band gap of 3.4 eV. It is a suitable material for applications in the optoelectronics industry [49] that range from light-emitting diodes to solar cells for space applications (satellites), as a consequence of its high stability in ionizing radiation ambients [50].

GaN p–n junctions, alloyed with indium (InGaN) and aluminum (AlGaN), facilitate the fabrication of light-emitting diodes that cover the visible spectral range

Gallium Arsenide

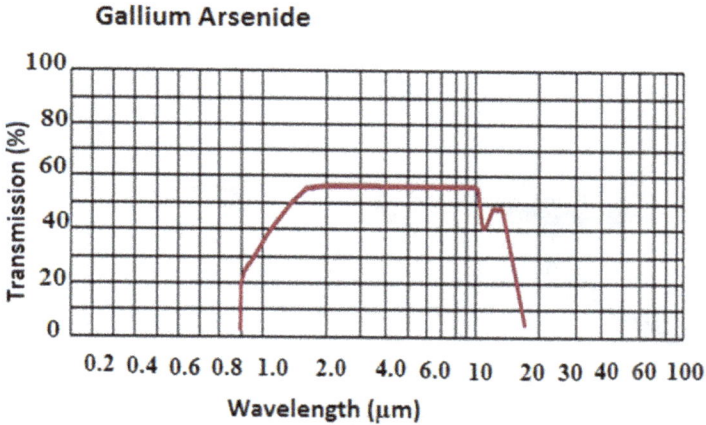

Figure 15.11. Gallium arsenide infrared transmission window band [46].

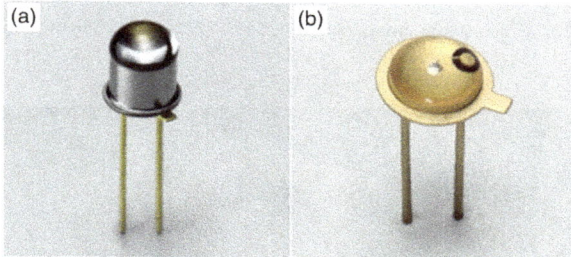

Figure 15.12. (a) GaAs infrared LED and (b) AlGaInP red emitting diode [47, 48].

from ultraviolet to red [51, 52]; these include the blue/UV-light-emitting diodes [53, 54], violet laser diodes to read blue-rays discs [55] and full color light-emitting diode displays.

OSRAM Optical Semiconductors offers a variety of visible laser diodes based on InGaN that are suited for applications in visual projection, stage lighting, metrology and spectroscopy. Figure 15.13 illustrates a green single mode laser, PL 520, in a TO38icut package manufactured by OSRAM OS [56].

Nichia Corporation, a light-emitting diode manufacturer, produces UV-LEDs of high quality that are required for high precision curing, ink curing (printing), and bill checker applications. Figure 15.14 shows a picture of a UV-LED with emissivity wavelengths of 365 nm, 385 nm, 395 nm and 405 nm [57].

Soraa, founded by Nobel Laureate in Physics, Shuji Nakamura, is the leading GaN-on-GaN LED manufacturer. Stressed GaN-on-GaN LEDs are capable of creating the perfect LED light source. The company's violet-emitting LEDs are made from GaN substrates. This technology combines GaN semiconductor materials with GaN substrates effectively lowering the lattice mismatch and resulting in a

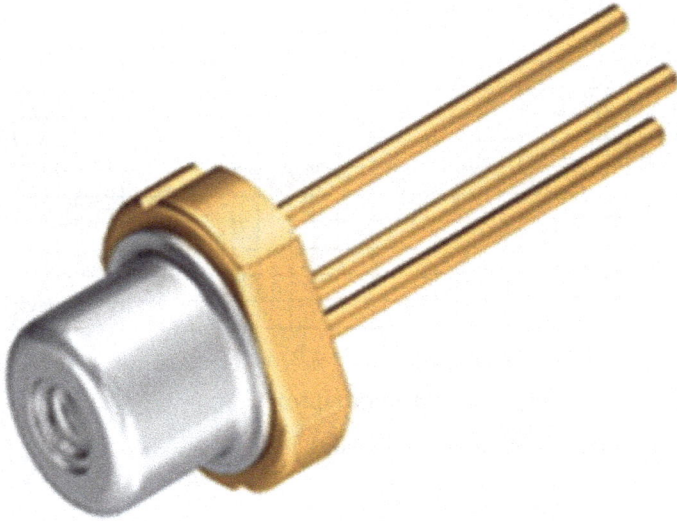

Figure 15.13. Green single mode laser diode PL 520, in a TO38icut package [56].

Figure 15.14. UV-LED model NVSU233A (Nichia Corporation) [57].

high radiation luminous efficacy [58]. Figure 15.15 shows a picture of a violet light-emitting LED chip using GaN-on-GaN technology manufactured by Soraa [58].

Samsung Electronics, a world leader in advanced component solutions, introduced a full line-up of GaN-on-Si wafer chip-scale LED packages (CSP). Combining advanced flip chip technology with phosphor coating technology, Samsung's CSP technology significantly scales down the size of a conventional LED package by eliminating any need for metal wires and plastic molds. The introduction of this new combination allows the manufacturing of more flexible and

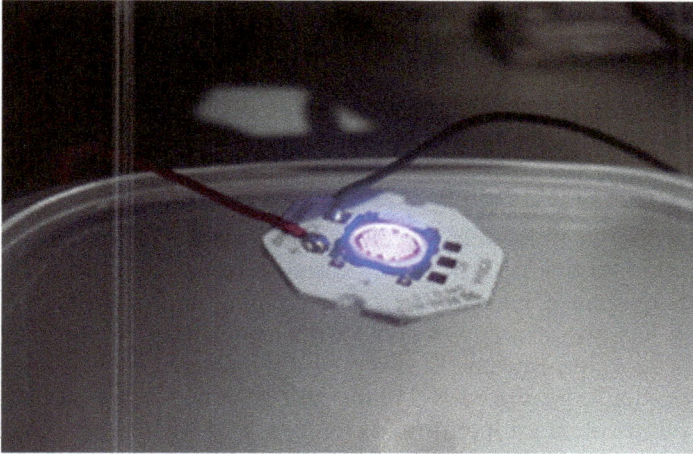

Figure 15.15. A violet light emitting LED chip using GaN-on-GaN technology (Soraa) [58].

Table 15.1. Samsung's chip-scale LED packages (CSP) [59].

Products	Light efficacy	Production schedule	Application
LM101A	149 lm W^{-1}	Mass production in January 2016	Bulbs, High Bay, Spot, D/L
LM102A	135 lm W^{-1}	Mass production in March 2016	Bulb, High Bay, Spot, D/L
LH181A	162 lm W^{-1}	Q2 2016	Street, Tunnel, High Bay
LH204A LH309A	124 lm W^{-1}, 127 lm W^{-1}	Q2 2016	MR, PAR, D/L
PoW	8 in Wafer technology	Q4 2016	All

compact designs of LED lighting modules or fixtures, lowering the production and operational costs of manufacturing of LED lighting systems [59]. Table 15.1 shows a list of the new technology of Samsung's chip-scale LED packages (CSP) and its applications [59].

15.7 Indium antimonide

Indium antimonide (InSb) is a compound semiconductor with a narrow energy band-gap of 0.17 eV at 300 K and 0.23 at 80 K [60], making it suitable to fabricate mid-infrared optical detectors [61]. InSb is commonly used in the fabrication of infrared detectors (thermal imaging cameras, FLIR systems and infrared astronomy). Another important application of InSb is as a terahertz (THz) radiation source as it is a strong photo-Dember emitter.

Infrared detectors made of InSb operate between 1–5 μm wavelengths. The quantum efficiency of InSb focal planes, prior to the application of an antireflection (AR) coating, is about 60% in the short wavelength infrared (SWIR) region of 1–5 μm. It is improved to higher than 90% with the application of AR coating, as shown in figure 15.16 [62]. InSb detectors are extremely sensitive to ambiance. However, it is required to add a complex cooling system that operates at cryogenic temperatures (~80 K), in order to reduce noise due to dark current introduced by its narrow band-gap [63].

Sofradir Group is a leading developer and manufacturer of InSb MWIR (medium wavelength infrared; 3–5 μm) based-infrared sensors for high-performance infrared imaging solutions for military, space and industrial applications. This company has an infrared product portfolio that covers the entire spectrum from visible to very far infrared. The stable InSb is the material of choice due to its excellent properties to produce focal plane arrays with a very high operability, low NETD (noise equivalent temperature difference) and with very high gain stability at 77 K [64]. Figure 15.17 shows the Sofradir state of the art InSb MWIR focal

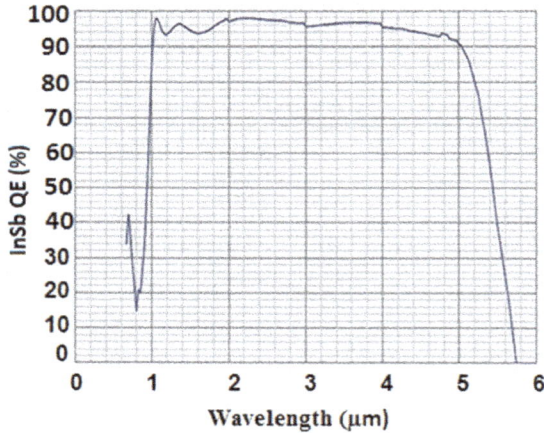

Figure 15.16. Typical spectral response of InSb FPA with antireflective coating for <1 to >5 μm [62].

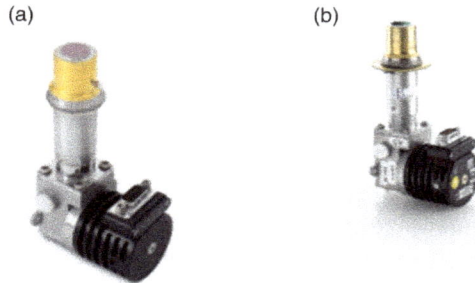

(a)   (b)

Figure 15.17. MWIR InSb Focal Plane arrays: a) Model INSPIR MW 384 × 288—25 μm pitch and b) Model AXIR MW 640 × 512—15 μm pitch [65, 66].

plane array (FPA) cryogenic detectors with high-speed and low-noise: (a) Model INSPIR MW 384 × 288—25 µm pitch—InSb and (b) Model AXIR MW 640 × 512—15 µm pitch—InSb. Both examples have a high stability and high performance with a detector spectral response of 3.7–4.8 µm and FPA operating temperature up to 8 K [65, 66].

Terahertz sensing science and technology has attracted the attention of InSb for applications in the fields of biochemistry and medicine. The vibrational or rotational energy transitions of the molecules, in the THz frequency range, allows their identification. This technique has applications in the field of sensing, imaging, tomography and spectroscopy [67–69].

InSb is a high efficiency emitter of terahertz radiation. This is due to the ultrafast buildup of the photo-Dember field in InSb [70]. It leads to applications of InSb in THz plasmonic devices [71].

15.8 Indium phosphide

Indium phosphide (InP) is a direct band-gap semiconductor of significant importance for its wide band-gap and its high electron mobility that makes it suitable for a wide variety of applications in high frequency optoelectronic devices [72–75] (diode lasers and fiber communication) and in the photovoltaics industry.

InP exhibits stable thermal characteristics, higher frequency response and low threshold voltage. It is one of the most efficient power devices for applications in fiber-optic telecommunications operating in the range of 1.3–1.55 µm, micro-wave photonics [76] and millimeter-wave frequency systems [77].

Most recently, InP diode lasers have attracted attention in the development of tunable diode laser spectroscopy (TDLS) in the near infrared spectrum: 1.26–1.67 µm. TDLS technique is considered as an important tool to detect the concentration of atmospheric gases (H_2O, CO, CO_2, CH_4 and NH_3) and monitor global warming and air pollution [78].

In the solar cell industry, InP mono-junction and multi-junction cells have achieved high efficiencies beyond 20% [79]. The cost of crystalline InP is higher than that of Si and GaAs. InP exhibits a higher tolerance to radiation damage. This facilitates its applications in solar cell arrays for use in space by agencies such as NASA, ESA, commercial and military sector for orbits with exposure to high radiation [80–82].

Princeton Lightwave, an optical imaging company based in Princeton, NJ, is the leading supplier of products based on Geiger-mode LiDAR technology. This technology is based on the InP detection and processing of single photons in real-time. The company develops and manufactures InP detectors that operate at near-infrared wavelengths with applications in imaging, autonomous vehicle navigation systems (Geiger Cruizer system) for driverless cars and optical communications (diode lasers and detectors). Most commercialized products include the Geiger cruizer, Geiger mode cameras, Geiger mode detectors and components (avalanche photodiodes, receivers and multimode laser diodes) [83].

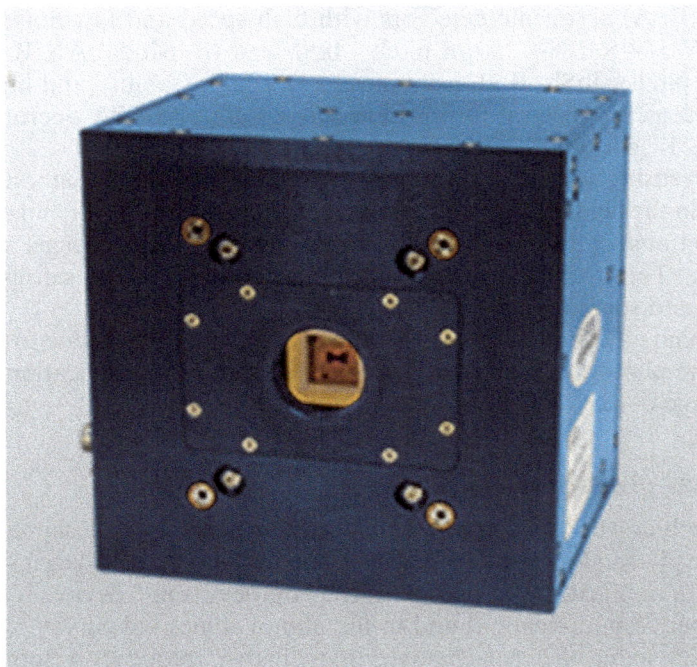

Figure 15.18. Merlin 32 × 32 free-running LiDAR camera [84].

The Princeton Lightwave asynchronous 32 × 32 Geiger-mode avalanche photo-diode (GmAPD) is based on flip-chip bonding of InP/InGaAsP GmAPD detector array. It is shown in figure 15.18 [84]. This camera contains a focal plane array (FPA) with hexagonal geometry pixels on a 65.8 µm pixel pitch formatted in a rectangular imaging area of 2.1 mm × 1.8 mm. Each pixel provides time-of-flight information with free-running operation and fully asynchronous timestamp readout. This camera captures images at accuracies and speeds that are orders of magnitude above that of other near-infrared imaging technologies. These images are used to create high resolution maps for the US Geological Survey, tracking satellites in geosynchronous orbits and also in airborne defense applications [84].

The National Renewable Energy Laboratory (NREL) has been focusing on the development of advanced multijunction solar cell technology and is the inventor of the original gallium indium phosphide/gallium arsenide (GaInP/GaAs) multi-junction cell with a high efficiency. NREL has demonstrated ∼46% efficiency with a four-junction IMM (Inverted Metamorphic Multijunction) solar cell. NREL is also working on the development of the concept of five- and six-junction IMM solar cells that are capable of exceeding 50% efficiency at high solar concentrations. Figure 15.19 shows the schematic of a target three-junction IMM solar cell structure developed by NREL [85].

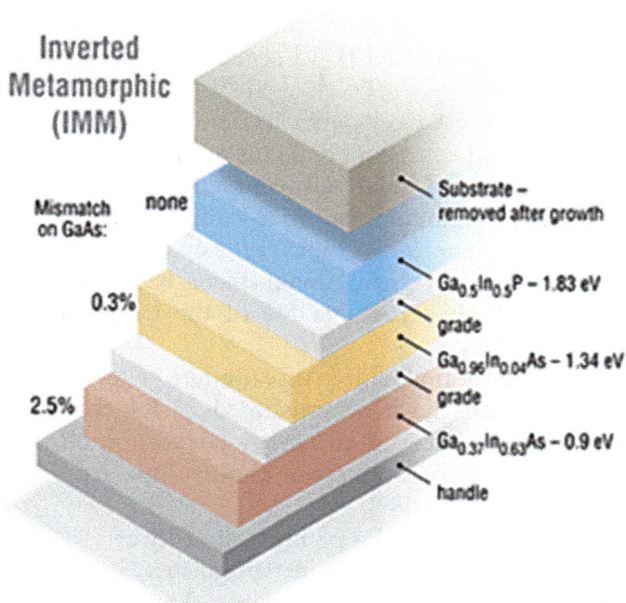

Figure 15.19. High-solar concentration III–V multijunction solar cells [85].

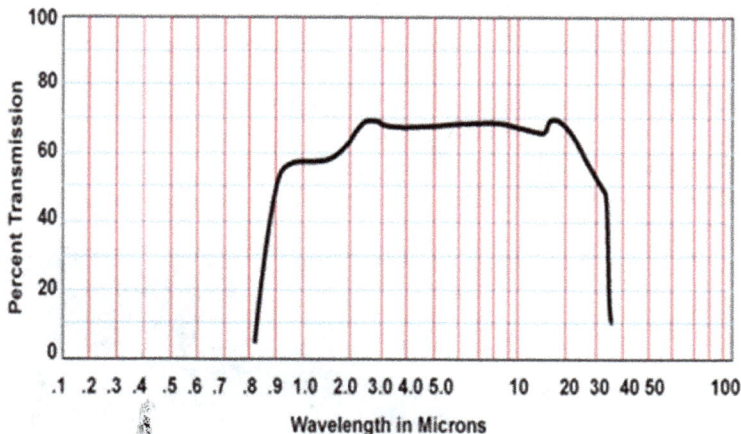

Figure 15.20. Cadmium telluride infrared transmittance window band [88].

15.9 Cadmium telluride

Cadmium telluride (CdTe) is a binary compound semiconductor with a direct band-gap of 1.5 eV at 300 K [86]. This compound semiconductor is widely used in several applications. These include the following: manufacture of CdTe solar cells, x-ray, gamma and beta ray-detectors, infrared optical windows and lenses [87].

Cadmium telluride is transparent from the infrared wavelength range of 0.83 μm to beyond 20 μm [88], as shown in figure 15.20. CdTe has a high resistance to moisture and can operate at elevated temperatures without dissociation. At the same time, it is also the softest among the II–VI semiconductors and is easily prone to scratching and cleaving.

Cadmium telluride based solar cells offer flexible thin film photovoltaic technology with high efficiency and low cost manufacturing, transportation and installation when compared to the standard solar cell technology based on c-Si. Solar panels made with single-junction cadmium telluride–cadmium sulfide (CdTe–CdS) are extremely thin and offer a significantly higher efficiency at higher temperatures and lower levels of illumination than conventional silicon-based solar panels. Cadmium telluride solar cell technology represents 5% of the worldwide solar cell power market.

First Solar is the major manufacturer of cadmium telluride solar panels in the United States. Research labs have demonstrated 22.1% conversion efficiency of CdTe cells. First Solar has achieved an energy conversion efficiency of 16.4% in commercial solar panels [89]. A commercial photovoltaic array made of cadmium telluride is shown in figure 15.21 [89].

First Solar, with more than 10 solar plants installed worldwide, generates energy in the order of gigawatts. It is a leader in the construction and operation of the largest grid-connected photovoltaic power plants. Topaz Solar Farm, one of the world's largest solar farms, is a 550 megawatt photovoltaics plant in San Luis Obispo County, California. This plant includes 9 million CdTe photovoltaic modules, based on thin-film technology, manufactured by First Solar [90].

Cadmium telluride has been studied and manufactured as an x-ray and gamma ray detector material over the last five decades. CdTe detectors have been developed to detect nuclear radiation with applications in a variety of fields including nuclear physics, x-ray, beta and gamma ray astronomy and nuclear medicine.

CdTe exhibits wide band-gap at room temperature in comparison to Si and Ge. Thus, this allows it to be considered as a promising material to fabricate radiation

Figure 15.21. Photovoltaic array made of cadmium telluride (First Solar) [89].

detectors with a high energy resolution and high detection efficiency at room temperature [91].

Eurorad produces semiconductor radiation detectors (pixels) and systems with applications in a variety of fields such as research, industry, medicine and the environment. The CdTe detectors, manufactured by this company, detect x, gamma, beta, neutron, and thermal radiation when operating at room temperature. Figure 15.22 shows the following: (a) CdTe radiation counter (Type: C. 10. 10. 10) for applications in x-ray diffraction, x-ray fluorescence, bone densitometry, surgical probes, custom inspection systems and radiation monitoring for nuclear safeguard; and (b) CdTe hemispheric detector model Hemi 500, a gamma-ray detector of hemispherical structure that exhibits good energy resolution even at 662 keV [92].

Dectris Ltd is a company specializing in the manufacture of CdTe radiation counters for application in the laboratory and industry. The company offers PILATUS3 X CdTe, a single-photon counting detector offering highly efficient detection up to 100 KeV with the developed technology for hard x-ray detection applications [93]. The PILATUS3 X CdTe detector systems are used to measure x-ray diffraction, high-energy x-ray diffraction (HE-XRD), x-ray diffraction tomography and microscopy (XRD-CT/XDM), x-ray powder diffraction, pair-distribution function analysis (XPD, PDF), imaging (x-ray projection imaging (radiography), x-ray computed tomography (CT) and x-ray phase-contrast imaging (PCI) [93]. Figure 15.23 shows a picture of PILATUS3 X CdTe systems—Hard x-ray detectors, manufactured by Dectris Ltd [93].

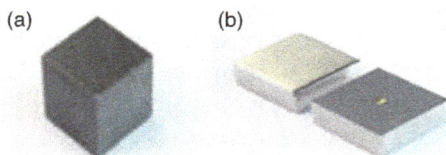

Figure 15.22. (a) CdTe counter and (b) CdTe hemispheric detectors [92].

Figure 15.23. PILATUS3 X CdTe systems—hard x-ray detectors (DECTRIS Ltd) [93].

15.10 Mercury cadmium telluride

Mercury cadmium telluride (HgCdTe) is the most important semiconductor material for infrared detection applications. HgCdTe was first developed for military applications (scanning systems) and later expanded for use by the commercial imaging industry (staring systems). HgCdTe photodiode, like silicon photodiodes, operates as a photovoltaic or photoconductive device [94]. The tunability of mercury cadmium telluride bandgap provides the capability to use this material as an absorber over the entire infrared spectral band [94]. The band gap energy tunability allows the infrared radiation detection in the following windows: short-wave (1–3 μm), mid-wave (3–5 μm), long-wave (8–14 μm). The band gap of mercury cadmium telluride, as a function of cadmium composition [95], is shown in figure 15.24. HgCdTe has large optical absorption coefficients that enable high quantum efficiency and an inherent recombination mechanism with the advantage of high operating temperatures [96]. These properties make mercury cadmium telluride an ideal material for the fabrication of infrared detectors.

The major limitation of HgCdTe-based LWIR detectors is that they need to operate at temperatures close to liquid nitrogen (77 K), to reduce noise due to thermally excited carriers. HgCdTe cameras, in the MWIR range, can be operated at temperatures that are accessible to thermoelectric coolers with a conservative loss in performance. The cooling system requires maintenance and adds weight to the detector system.

HgCdTe-based infrared detectors have been developed with the introduction of new microfabrication techniques in combination with advances in the opto-electronic systems. HgCdTe-based infrared detectors can be classified into three generations. The first generation—linear arrays of photoconductive detectors, the second generation—2-D arrays of photovoltaic detectors and the third generation—avalanches photodiodes and two-color detectors.

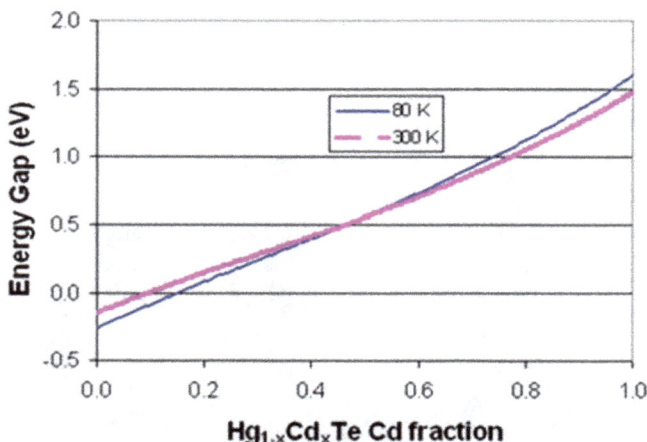

Figure 15.24. Mercury cadmium telluride bandgap as a function of cadmium composition [95].

15-19

The first generation HgCdTe-based infrared detectors have been developed in linear arrays of photoconductive devices. This does not include multiplexing functions in the focal plane and the image is obtained by scanning the scene across the line array strip. The first system to use this technology is the US army common module HgCdTe array having 60, 120 and 180 elements, with element pixel of 50 μm [96]. An example of the 180 element array is presented in figure 15.25 [96].

The second generation HgCdTe infrared detectors are configured in two-dimensional arrays and produced in scanning and staring formats. The integration of more detector elements in both formats improves the sensitivity and the spatial resolution of the second generation detectors. Figure 15.26 shows an example of the second generation photovoltaic detectors which are configured in scanning and staring formats [97].

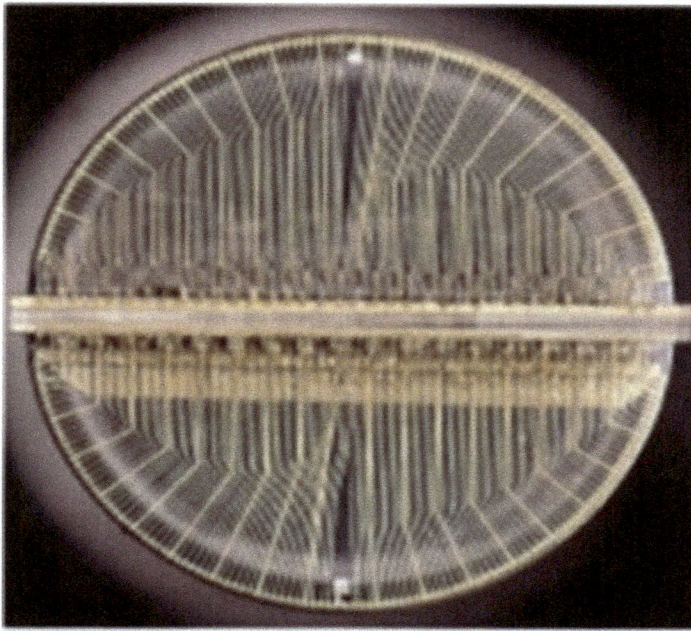

Figure 15.25. First generation 180 elements common module linear array of HgCdTe photoconductive detectors [96].

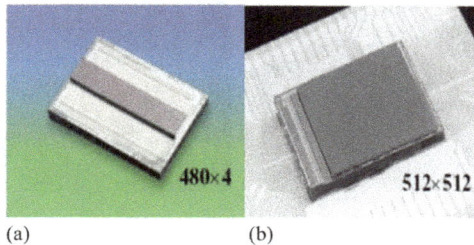

(a)   (b)

Figure 15.26. Second generation PV HgCdTe-based infrared detectors which are configured in both scanning (a) and staring (b) formats [97].

In third generation HgCdTe infrared detectors, the performance of the system is enhanced with the increase of the array size, readout capability, two-color capacity and improvement of cooling device. The two-color infrared detector systems detect infrared radiation in two separated infrared spectral bands: first, detect an object and then identify it. Multicolor capabilities are crucial for advanced recognition infrared imaging systems as it provides image contrast target identification. Raytheon Vision Systems has developed a two-color, 1280 × 720 large formats with 20 × 20 μm unit pixels. Figure 15.27 shows the infrared array incorporated with the ROIC (Read out integrated circuit) [98]. Figure 15.28 illustrates an example of a two-color imagery system from one of over 70 arrays in the 480 × 640 format built by Raytheon for a double band: mid-wave infrared (MWIR) and long-wave infrared (LWIR) [99].

Table 15.2 presents a compilation of representative infrared focal plane array (FPA) imaging systems that are commercially available as standard products from major manufacturers [100].

Figure 15.27. Third generation dual-band megapixel MW/LW FPAs: RVS 1280_720 format HgCdTe FPAs mounted on Dewar platforms [98].

Figure 15.28. Example of imagery obtained at 78 K for an Army 640_480 M/LWIR FPA (#7586704) at *f*/5 field of view and 30 Hz frame rate. The subject is holding a thin piece of plastic which transmits in the MWIR band but absorbs in the LWIR band [99].

Table 15.2. Representative IR FPAs offered by some major manufacturers [100].

Manufacturer/Web site	Size/Architecture	Pixel size (μm)	Detector material	Spectral range (μm)	Oper. Temp. (K)	$D^*(\lambda p)$ (cmHz$^{1/2}$ W^{-1})/ NETD (mK)
Goodrich Corporation	320 × 240/H	25 × 25	InGaAs	0.9–1.7	300	1 × 10^{13}
www.Sensorsinc.com	640 × 512/H	25 × 25	InGaAs	0.4–1.7	300	>6 × 10^{12}
Raytheon Vision Systems	1024 × 1024/H	30 × 30	InSb	0.6–5.0	50	
www.raytheon.com/businesses/ ncs/rvs/index.html	2048 × 2048/H (Orion II)	25 × 25	HgCdTe	0.6–5.0	32	
	2048 × 2048/H (Virgo-2k)	20 × 20	HgCdTe	0.8–2.5	4–10	23
	2048 × 2048/H	15 × 15	HgCdTe/Si	3.0–5.0	78	
	1024 × 1024/H	25 × 25	Si:As	5–28	6.7	
	2048 × 1024/H	25 × 25	Si:As	5–28		
Teledyne Imaging Sensors	4096 × 4096/H (H4RG)	10 × 10 or 15 × 15	HgCdTe	1.0–1.7	120	
http://teledynesi.com/imaging/	4096 × 4096/H (H4RG)	10 × 10 or 15 × 15	HgCdTe	1.0–2.5	77	
	4096 × 4096/H (H4RG)	10 × 1 0 or 15 × 15	HgCdTe	1.0–5.4	37	
	2048 × 2048/H (H2RG)	18 × 18	HgCdTe	1.0–1.7	120	
	2048 × 2048/H (H2RG)	18 × 18	HgCdTe	1.0–2.5	77	
	2048 × 2048/H (H2RG)	18 × 18	HgCdTe	1.0–5.4	37	
Sofradir	1000 × 256/H (Saturn)	30 × 30	HgCdTe	0.8–2.5	≤200	
www.sofradir.com/	1280 × 1024/H (Jupiter)	15 × 15	HgCdTe	3.7–4.8	77–110	18

Company	Format	Pixel size	Material	Spectral range (µm)	Temperature (K)	
	384 × 288/H (Venus)	25 × 25	HgCdTe	7.7–9.5	77–80	17
	640 × 512/H	20 × 20	QWIP	8.0–9.0	73	31
	640 × 512/H	24 × 24	HgCdTe	MW(dual)	77–80	15–20
	640 × 512/H	24 × 24	HgCdTe	MW/LW(dual)	77–80	20–25
Selex	124 × 768/H (Merlin)	16 × 16	HgCdTe	3–5	Up to 140	15
www.selexsas.com/	640 × 512/H (Eagle)	24 × 24	HgCdTe	8–10	Up to 90	24
SelexGalileo/En//index.sdo	640 × 512/H (Condor)	24 × 24	HgCdTe	MW/LW(dual)	80	28/28
AIM	640 × 512/H	24 × 24	HgCdTe	3–5		25
www.aim-ir.com	640 × 512/H	15 × 15	HgCdTe	8–9		40
	384 × 288	40 × 40	Type II SL	MW(dual)		35/25
SCD	1280 × 1024/H	15 × 15	InSb	3–5	77	20
www.scd.co.il						
DRS Technologies	2048 × 2048/H	18 × 18	Si:As	5–28	7.8	
www.drsinfrared.com	1024 × 1024/H	25 × 25	Si:As	5–28	7.8	
	2048 × 2048/H	18 × 18	Si:Sb	5–40	7.8	

H—hybrid, M—monolithic

List of Patents for semiconductors discussed in the book—with emphasis on optical/optoelectronic devices

Silicon

Patent title	Patent number/ Pub. number	Date of patent/ pub. date
Thin film solar cells on flexible substrates and methods of constructing the same	US 9,590,133 B1	Mar. 7, 2017
Solar cell assembly II	US 9,590,126 B2	Mar. 7, 2017
Light concentrator and a solar cell	US 9,595,912 B2	Mar. 14, 2017
Systems and methods for cascading photovoltaic structures	US 9,590,132 B2	Mar. 7, 2017
Photovoltaic system	US 9,583,645 B2	Feb. 28, 2017
Photovoltaic cells withmulti-band gap and applications in a low temperature polycrystalline silicon thin film transistor panel	US 9,577,137 B2	Feb. 21, 2017
Solar cell emitter region fabrication using self-aligned implant and cap	US 9,577,134 B2	Feb. 21, 2017
Vision system for vehicle	US 9,609,289 B2	Mar. 28, 2017
Devices, methods, and systems for expanded-field-of-view image and video capture	US 9,609,190 B2	Mar. 28, 2017
Mobile device protection case	US 9,609,101 B1	Mar. 28, 2017
Image pickup apparatus	US 9,609,188 B2	Mar. 28, 2017

Germanium

Patent title	Patent number/ Pub. number	Date of patent/ pub. date
Germanium metal-contact-free near-IR photodetector	US 9,553,222 B2	Jan. 24, 2017
Solar cell and solar cell assembly	US 9,508,876 B2	Nov. 29, 2016
Solar cell having doped buffer layer and method of fabricating the solar cell	US 9,520,530 B2	Dec. 13, 2016
Multi-junction solar cell	US 9,530,921 B2	Dec. 27, 2016
Solar cell having doped semiconductor heterojunction contacts	US 9,548,409 B2	Jan. 17, 2017
Light-emitting device and method for manufacturing the same	US 9,196,813 B2	Nov. 24, 2015
Photoreceptor with improved blocking layer	US 9,123,842 B2	Sep. 1, 2015
Photodetector focal plane array systems and methods	US 9,362,324 B1	Jun. 7, 2016
Monolithic visible-infrared focal plane array on silicon	US 9,472,588 B1	Oct. 18, 2016
Two color detector leveraging resonant cavity enhancement for performance improvement	US 9,536,917 B2	Jan. 3, 2017

Graphene

Patent title	Patent number/ Pub. number	Date of patent/ pub. date
Adjustable wearable system having a modular sensor platform	US 9,592,007 B2	Mar. 14, 2017
High-density 3D graphene-based monolith and related materials, methods, and devices	US 9,601,226 B2	Mar. 21, 2017
Light-emitting element	US 9,604,928 B2	Mar. 28, 2017
Display screen and its manufacturing process	US 9,606,378 B2	Mar. 28, 2017
Semiconductor device, electronic component, and electronic device	US 9,607,569 B2	Mar. 28, 2017
Deformable display device andmethod for controlling thereof	US 9,606,648 B2	Mar. 28, 2017
Touch panel and manufacturing method thereof and touch display panel	US 9,606,679 B2	Mar. 28, 2017
Touch sensing apparatus and touchscreen apparatus including the same	US 9,606,688 B2	Mar. 28, 2017
Light-emitting diode	US 9,608,168 B2	Mar. 28, 2017
Process for the preparation of graphene	US 8,715,610 B2	May 6, 2014
Support structures for an attachable, two-dimensional flexible electronic device	US 9,560,751 B2	Jan. 31, 2017

Silicon carbide

Patent title	Patent number/ Pub. number	Date of patent/ pub. date
Multi-segment led components and led lighting apparatus including the same	US 9,609,709 B2	Mar. 28, 2017
Light-emitting diode module having light-emitting diode joined through solder paste and light-emitting diode	US 9,608,186 B2	Mar. 28, 2017
Light-emitting device	US 9,608,180 B2	Mar. 28, 2017
Light-emitting device package and backlight unit having the same	US 9,608,177 B2	Mar. 28, 2017
Light-emitting diode and application therefor	US 9,608,171 B2	Mar. 28, 2017
Light-emitting diode	US 9,608,168 B2	Mar. 28, 2017
Method of manufacturing light emitting element	US 9,608,170 B2	Mar. 28, 2017
Light emitting device	US 9,608,167 B2	Mar. 28, 2017
Light emitting diode and method of fabricating the same	US 9,608,165 B2	Mar. 28, 2017
High radiance light emitting diode light engine	US 9,606,304 B2	Mar. 28, 2017
Light-emitting diode (LED) components including led dies that are directly attached to lead frames	US 9,601,673 B2	Mar. 21, 2017
High brightness led package	US 9,601,672 B2	Mar. 21, 2017

Gallium arsenide

Patent title	Patent number/ Pub. number	Date of patent/ pub. date
6.1 angstrom III–V and II–VI semiconductor platform	US 8,891,573 B2	Nov. 18, 2014
Light-emitting devices and substrates with improved plating	US 9,590,155 B2	Mar. 7, 2017
Light-emitting module	US 9,590,139 B1	Mar. 7, 2017
Light-emitting device and method of manufacturing the same	US 9,543,475 B2	Jan. 10, 2017
Strain engineered bandgaps	US 9,595,624 B2	Mar. 14, 2017
III–V photonic integrated circuits on silicon substrate	US 9,595,805 B2	Mar. 14, 2017
Photovoltaics on silicon	US 9,508,890 B2	Nov. 29, 2016
Method for forming ohmic n-contacts at low temperature in inverted metamorphic multijunction solar cells with contaminant isolation	US 9,287,438 B1	Mar. 15, 2016
Ohmic n-contact formed at low temperature in inverted metamorphic multijunction solar cells	US 8,987,042 B2	Mar. 24, 2015
Gallium arsenide solar cell with germanium/palladium contact	US 8,753,918 B2	Jun. 17, 2014
Light-emitting device and camera-equipped cellular phone incorporating the same	US 7,938,550 B2	May 10, 2011

Gallium nitride

Patent title	Patent number/ Pub. number	Date of patent/ pub. date
Light-emitting device (LED) array unit and led module comprising the same	US 9,603,214 B2	Mar. 21, 2017
Low voltage laser diodes on {20–21} gallium and nitrogen containing substrates	US 9,543,738 B2	Jan. 10, 2017
Color stable red-emitting phosphors	US 9,580,648 B2	Feb. 28, 2017
Laminated display unit	US 9,599,766 B2	Mar. 21, 2017
Solid state lighting devices without converter materials and associated methods of manufacturing	US 9,601,658 B2	Mar. 21, 2017
Structure for a gallium nitride (GaN) high electron mobility transistor	US 9,601,608 B2	Mar. 21, 2017
Semiconductor light-emitting device, nitride semiconductor layer, and method for forming nitride semiconductor layer	US 9,601,662 B2	Mar. 21, 2017
Bipolar junction transistor with improved avalanche capability	US 9,601,605 B2	Mar. 21, 2017
GaN-on-Si switch device	US 9,601,638 B2	Mar. 21, 2017
Method of producing group III nitride semiconductor light-emitting device	US 9,601,654 B2	Mar. 21, 2017

Light-emitting diode chip	US 9,601,663 B2	Mar. 21, 2017
Light-emitting device	US 9,601,667 B2	Mar. 21, 2017
Method of manufacturing low cost, high efficiency LED	US 9,601,656 B1	Mar. 21, 2017

Indium antimonide

Patent title	Patent number/ Pub. number	Date of patent/ pub. date
Nano avalanche photodiode architecture for photon detection	US 9,570,646 B2	Feb. 14, 2017
Light-emitting diodes	US 9,570,652 B2	Feb. 14, 2017
Thin-film coatings, electro-optic elements and assemblies incorporating these elements	US 9,529,214 B2	Dec. 27, 2016
Two color detector leveraging resonant cavity enhancement for performance improvement	US 9,536,917 B2	Jan. 3, 2017
MEMS-released curved image sensor	US 9,551,856 B2	Jan. 24, 2017
Spectral imaging system for remote and noninvasive detection of target substances using spectral filter arrays and image capture arrays	US 9,551,616 B2	Jan. 24, 2017
Display device	US 9,564,608 B2	Feb. 7, 2017
Light-emitting devices with optical elements and bonding layers	US 9,583,683 B2	*Feb. 28, 2017
LED display with wavelength conversion layer	US 9,599,857 B2	Mar. 21, 2017
Photoelectric conversion element	US 9,601,643 B2	Mar. 21, 2017

Indium phosphide

Patent title	Patent number/ Pub. number	Date of patent/ pub. date
Phosphor optical element and light-emitting device using the same	US 9,605,832 B2	Mar. 28, 2017
Light-emitting device having a plurality of contact parts	US 9,577,170 B2	Feb. 21, 2017
LED display with wavelength conversion layer	US 9,599,857 B2	Mar. 21, 2017
Light-emitting device and light emitting device package	US 9,595,647 B2	*Mar. 14, 2017
Full spectrum light-emitting arrangement	US 9,599,293 B2	Mar. 21, 2017
Nanostructured LED	US 9,595,649 B2	Mar. 14, 2017
Backlight unit and liquid crystal display device including the same	US 9,606,281 B2	Mar. 28, 2017
Light-emitting device package and backlight unit having the same	US 9,608,177 B2	Mar. 28, 2017

(*Continued*)

LED structures for reduced non-radiative sidewall recombination	US 9,601,659 B2	Mar. 21, 2017
Light-emitting device	US 9,601,667 B2	Mar. 21, 2017
Optical component, products including same, and methods for making same	US 9,605,833 B2	Mar. 28, 2017
Semiconductor device	US 9,613,999 B2	Apr. 4, 2017
Semiconductor devices with a thermally conductive layer	US 9,614,046 B2	Apr. 4, 2017

Cadmium telluride

Patent title	Patent number/ Pub. number	Date of patent/ pub. date
Photovoltaic module	US 9,602,048 B2	Mar. 21, 2017
Radiation detection	US 9,581,619 B2	Feb. 28, 2017
Photovoltaic cells withmulti-band gap and applications in a low temperature polycrystalline silicon thin film transistor panel	US 9,577,137 B2	Feb. 21, 2017
Semiconductor device with different fin pitches	US 9,589,956 B1	Mar. 7, 2017
Thin film solar cells on flexible substrates and methods of constructing the same	US 9,590,133 B1	Mar. 7, 2017
Photo-induced phase transfer of luminescent quantum dots	US 9,598,635 B2	Mar. 21, 2017
Counting digital x-ray detector and method for recording an x-ray image	US 9,599,730 B2	Mar. 21, 2017
Photoelectric conversion element	US 9,601,643 B2	Mar. 21, 2017
Solar cell module	US 9,601,646 B2	Mar. 21, 2017
Photovoltaic device	US 9,602,046 B2	Mar. 21, 2017

Mercury cadmium telluride

Patent title	Patent number/ Pub. number	Date of patent/ pub. date
Minority carrier based HgCdTe infrared detectors and arrays	US 8,686,471 B2	Apr. 1, 2014
Nano avalanche photodiode architecture for photon detection	US 9,570,646 B2	Feb. 14, 2017
Schottky barrier diode and apparatus using the same	US 9,553,211 B2	Jan. 24, 2017
Radiation detector having a band gap engineered absorber	US 9,541,450 B2	Jan. 10, 2017
Method of making an optical detector	US 9,520,525 B1	Dec. 13, 2016
Dual-band detector array	US 9,490,292 B1	Nov. 8, 2016
Mid-infrared photodetectors	US 9,318,628 B2	Apr. 19, 2016

References

[1] Mitchell B S 2004 *An Introduction to Materials Engineering for Chemical and Materials Engineers* (Hoboken, NJ: Wiley)

[2] Rudnitsky A, Agdarov S, Gulitsky K and Zalevsky Z 2017 Silicon based mechanic-photonic wavelength converter for infrared photodetection *Opt. Commun.* **392** 114–8

[3] Hummel R E 2011 *Electronic Properties of materials* (New York: Springer)

[4] Rollin B V and Simmons E L 1953 Long Wavelength infrared photoconductivity of Silicon at low temperatures Proc. Physical Soc. B **66** 162–8

[5] Sclar N 1984 Properties of doped silicon and germanium infrared detector *Prog. Quantum Electron.* **9** 149–257

[6] Rogalski A 2011 *Infrared Detectors* 2nd edn (Boca Raton, FL: CRC)

[7] Saga T 2010 Advances in crystalline silicon solar cell technology for industrial mass production *NPG Asia Mater.* **2** 96–102

[8] Lee Y, Park C, Balaji N, Lee Y-J and Dao V A 2015 High-efficiency silicon solar cells: a review *Isr. J. Chem.* **55** 1050– 63

[9] Chen Y M, Chiu Y and Wu H 2012 Improved testing of soldered interconnects quality on silicon solar cell *Glob. Perspect. Eng. Manage* **1** 51–8

[10] Li H H 1993 Refractive index of silicon and germanium and its wavelength and temperature derivatives *J. Phys. Chem. Ref. Data* **9** 561–658

[11] http://www.janostech.com/knowledge-center/ir-coating-information/what-are-the-specifications-of-a-silicon-si-2-5-ir-lens.html (accessed 5 March 2017)

[12] https://www.edmundoptics.com/optics/optical-lenses/plano-convex-pcx-spherical-singlet-lenses/standard-silicon-plano-convex-pcx-lenses/3187/ (accessed 5 March 2017)

[13] https://www.edmundoptics.com/optics/optical-lenses/aspheric-lenses/silicon-aspheric-lenses/3699/ (accessed 5 March 2017)

[14] Mashanovich G Z *et al* 2017 Germanium mid-infrared photonic devices *J. Lightwave Technol.* **35** 624–30

[15] Johnson L C, Campbell D L, Ovchinnikov O S and Peterson T E 2011 Performance characterization of a high-purity germanium detector for small-animal SPECT *IEEE Nuclear Science Symposium Conference Record IEEE* 607–12

[16] Fast J E, Aalseth C E, Caggiano J A, Day A R, Erin S, Fuller E S, Hossbach T W, Hyronimus B J, Runkle R and Warren G A 2009 A high-efficiency fieldable germanium detector array *IEEE Trans. on Nucl. Sci.* **56** 1224–8

[17] Rumaiz A K, Krings T, Siddons D P, Kuczewski A J, Protić D and Ross C 2014 De Geronimo G and Zhong Z 2014 A monolithic segmented germanium detector with highly integrated readout *IEEE Trans. on Nucl. Sci.* **61** 3721–6

[18] Colace L and Assanto G 2009 Germanium on silicon for near-infrared light sensing *IEEE Photon. J.* **1** 69–79

[19] Fernandez J, Janz S, Suwito D, Oliva E and Dimroth F 2008 Advanced concepts for high-efficiency germanium photovoltaic cells *Photovoltaic Specialists Conference, 2008. PVSC '08. 33rd IEEE (San Diego, CA, 11–16 May 2008)*

[20] Kim D, Choi Y, Do E C, Lee Y and Kim Y G 2012 High efficiency silicon and germanium stack junction solar cells *Electron Devices Meeting (IEDM), 2012 (San Francisco CA, 10–13 Dec 2012) IEEE* 295–8

[21] Chang P-S and Shen C-S 2007 A new fast infrared tracking system with thermopile array implementation *Int. Conf. on Optical MEMS and Nanophotonics, 2007 IEEE/LEOS (Hualien, Taiwan, 12 Aug–16 July 2007) IEEE* 97-8

[22] Moskalyk R R 2004 Review of germanium processing worldwide *Minerals Eng.* **17** 393–402

[23] Rieke G H 2007 Infrared detector arrays for astronomy *Ann. Rev. Astrophys.* **45** 77–115

[24] Brown R D Jr 2000 Germanium, U.S. Geological Survey (retrieved 2008-09-22)

[25] http://www.janostech.com/knowledge-center/ir-coating-information/what-are-the-speci fications-of-germanium-ge-8-12-ir-lens.html (accessed 5 March 2017)

[26] https://www.edmundoptics.com/optics/optical-lenses/aspheric-lenses/germanium-infrared-ir-aspheric-lenses/3335/ (accessed 5 March 2017)

[27] https://www.edmundoptics.com/optics/optical-lenses/achromatic-lenses/infrared-ir-achro matic-lenses/3224 (accessed 5 March 2017)

[28] Novoselov K S, Geim A K, Morozov S V, Jiang D, Zhang Y, Dubonos S V, Grigorieva I V and Firsov A A 2004 Electric field effect in atomically thin carbon films *Science* **306** 666–9

[29] Cooper D R *et al* 2012 Experimental review of graphene *ISRN Condens. Matter Phys.* **2012** 1–56

[30] Zhong M, Xu D, Yu X, Huang K, Liu X, Xu Y and Yang D 2016 Interface coupling in graphene/fluorographene heterostructure for high-performance graphene/silicon solar cells *Nano Energy* **28** 12–8

[31] Akinwande D, Tao L, Yu Q, Lou X, Peng P and Kuzum D 2015 Large-area graphene electrodes: using CVD to facilitate applications in commercial touchscreens, flexible nanoelectronics, and neural interfaces *IEEE Nanotechnol. Mag* **9** 6–14

[32] http://www.cz2dcarbon.com/ (accessed 9 March 2017)

[33] https://hubpages.com/technology/Whats-the-Latest-Technology (accessed 9 March 2017)

[34] Bhatnagar M and Baliga B J 1993 Comparison of 6H-SiC, 3C-SiC, and Si for power devices *IEEE Trans. Electron Devices* **40** 645–5

[35] https://www.digikey.com/en/articles/techzone/2016/dec/silicon-carbide-history-and-applications

[36] Stringfellow G B 1997 *High brightness light emitting diodes* (New York: Academic) pp 48, 57, 425

[37] http://www.cree.com/ (accessed 9 March 2017)

[38] http://www.cree.com/led-components/media/documents/XLampXPE2.pdf (accessed 9 March 2017)

[39] http://www.esa.int/ESA (accessed 9 March 2017)

[40] Petrovsky G T, Tolstoy M N, Ljubarsky S V, Khimitch Y P and Robb P N 1994 Proc. SPIE 2199, Advanced Technology Optical Telescopes V, 263

[41] http://sci.esa.int/herschel/ (accessed 15 March 2017)

[42] http://sci.esa.int/gaia/ (accessed 15 March 2017)

[43] Moon S, Kim K, Kim Y, Heo J and Lee J 2016 Highly efficient single-junction GaAs thin-film solar cell on flexible substrate *Sci. Rep.* **6** 30107

[44] http://www.altadevices.com/technology/ (accessed 7 March 2017)

[45] Skauli T *et al* 2003 Improved dispersion relations for GaAs and applications to nonlinear optics *J. Appl. Phys.* **94** 6447–55

[46] http://www.janostech.com/knowledge-center/optical-materials-guide/gallium-arsenide-gaas.html (accessed 7 March 2017)

[47] http://www.hamamatsu.com/us/en/L2388-01.html (accessed 7 March 2017)

[48] http://www.hamamatsu.com/us/en/L11767.html (accessed 7 March 2017)

[49] Di Carlo A 2001 Tuning optical properties of GaN-based nanostructures by charge screening *Phys Status Solidi* a **183** 81–5

[50] Arakawa Y 2002 Progress in GaN-based quantum dots for optoelectronics applications *IEEE J. Sel. Top. Quantum Electron.* **8** 823–32

[51] Lidow A, Witcher J B and Smalley K 2011 Enhancement mode gallium nitride (eGaN) FET characteristics under long term stress *GOMAC Tech Conference*

[52] Morkoç H, Strite S, Gao G B, Lin M E, Sverdlov B and Burns M 1994 Large-band-gap SiC, III–V nitride, and II–VI ZnSe-based semiconductor device technologies *J. Appl. Phys.* **76** 1363

[53] Amano H, Kito M, Hiramatsu K and Akasaki I 1989 P-Type conduction in Mg-doped GaN treated with low-energy electron beam irradiation (LEEBI) *Japan. J. Appl. Phys.* **28** L2112

[54] Nguyen X L, Nguyen T N N, Chau V T and Dang M C 2010 The fabrication of GaN-based light emitting diodes (LEDs) *Adv. Nat. Sci.: Nanosci. Nanotechnol.* **1** 025015

[55] DenBaars S P *et al* 2013 Development of gallium-nitride-based light-emitting diodes (LEDs) and laser diodes for energy-efficient lighting and displays *Acta Mater.* **61** 945–51

[56] http://www.osram-os.com/osram_os/en/products/product-catalog/laser-diodes/visible-laser/green-laser/index.jsp (accessed 7 March 2017)

[57] http://www.nichia.co.jp/en/product/uvled.html#NVSU233A (accessed 7 March 2017)

[58] http://www.ledinside.com/showreport/2016/3/top_led_technology_trends_at_light_building_2016 (accessed 7 March 2017)

[59] https://news.samsung.com/global/samsung-introduces-full-line-up-of-led-components-based-on-chip-scale-packaging-technology-based-on-chip-scale-packaging-technology

[60] http://www.ioffe.ru/SVA/NSM/Semicond/InSb/bandstr.html#Doping (accessed 7 March 2017)

[61] D'Costa V R, Kian K H, Jia B W, Yoon S F and Yeo Y-C 2015 Mid-infrared to ultraviolet optical properties of InSb grown on GaAs by molecular beam epitaxy *J. Appl. Phys.* **117** 223106

[62] https://www.photonics.com/Article.aspx?AID=58415 (accessed 9 March 2017)

[63] Chen H, Sun X, Lai K W C, Meyyappan M and Xi N 2009 Infrared Detection Using an InSb Nanowire *2009 IEEE Nanotechnology Materials and Devices Conf. (Traverse City, MI, 2–5 June, 2009)* 212–6

[64] Bai J, Weida Hu, Guo N, Lei W, Lv Y, Zhang X, SI J, Chen X and Lu W 2014 Performance optimization of InSb infrared focal-plane arrays with diffractive microlenses *J. Electron. Mater.* **43** 2795–801

[65] http://www.sofradir.com/product/inspir-mw-2/ (accessed 9 March 2017)

[66] http://www.sofradir.com/product/axir-mw/ (accessed 9 March 2017)

[67] Shourie Ranjana J, Bhatt P, Deshmukh P, Sangala B R, Satyanarayan M N, Umesh G and Prabhu S S 2015 *Indium antimonide (InSb) waveguide based THz sensor 2015 40th Int. Conf. on Infrared, Millimeter, and Terahertz waves (IRMMW-THz) (Hong Kong, 23-28 Aug. 2015)*

[68] Mittleman D M, Hunsche S, Boivin L and Nuss M C 1997 T-ray tomography *Opt. Lett.* **22** 904–6

[69] Mittleman D M, Jacobsen R H, Neelamani R, Baraniuk R G and Nuss M C 1998 Gas sensing using terahertz time-domain spectroscopy *Appl. Phys.* B **67** 379–90

[70] Gu P, Tani M, Kono S and Sakai K 2002 Study of terahertz radiation from InAs and InSb *J. Appl. Phys.* **91** 5533

[71] Tao J, Hu B, He X Y and Wang Q J 2013 Tunable subwavelength terahertz plasmonic stub waveguide filters *IEEE Trans. Nanotechnol.* **12** 1191–7

[72] Turley S 1984 Developments in indium-phosphide lasers *Electron. Power* **1984** 857–60

[73] Asar T, Ozcelik S and Ozbay E 2014 Structural and electrical characterizations of InxGa1-xAs/InP structures for infrared photodetector applications *J. Appl. Phys.* **115** 104502

[74] Gudiksen M S, Lauhon L J, Wang J, Smith D C and Lieber C M 2002 Growth of nanowire super lattice structures for nanoscale photonics and electronics *Nature* **415** 617–20

[75] Wang J F, Gudiksen M S, Duan X F, Cui Y and Lieber C M 2001 Highly polarized photoluminescence and photodetection from single indium phosphide nanowires *Science* **293** 1455–7

[76] Mathur A 2000 High-power indium phosphide semiconductor lasers *Lasers and Electro-Optics Society 2000 Annual Meeting. LEOS 2000 (Puerto Rico, 13–16 Nov 2000) IEEE*

[77] Szweda R 2003 InPb manufacturing remains on the cusp, III–Vs Review *Adv. Semicond. Mag.* **16** 36–40 May 2003

[78] Yoshimura R, Oohashi H, Sato T, Mitsuhara M and Kohtoku M 2014 Tunable diode laser spectroscopy (TDLS) and laser light sources *2014 IEEE Photonics Society Summer Topical Meeting Series (Montreal, 14–16 July 2014)*

[79] Jain R K and Flood D J 1991 Influence of the dislocation density on the performance of heteroepitaxial indium phosphide solar cells *IEEE Trans. Electron Devices* **40** 1928–34

[80] Goradia C, Thesling W and Goradia M G 1988 Predicted performance of near-optimally designed indium phosphide space solar cells at high intensities and temperatures *IEEE* 695–8

[81] Parrott J E and Potts A Indium phosphide solar cells: P+-N or N+-P? *22nd Photovoltaic Specialists Conf. (Las Vegas, NV, 7–11 Oct. 1991)* 153–8

[82] Jain R K and Flood D J 1994 Simulation of high-efficiency n+p indium phosphide solar cell results and future improvements *IEEE Trans. Electron Devices* **41** 2473–5

[83] https://www.princetonlightwave.com/products/ (accessed 15 March 2017)

[84] https://www.princetonlightwave.com/products/geiger-mode-cameras/ (accessed 15 March 2017)

[85] https://www.nrel.gov/pv/high-concentration-iii-v-multijunction-solar-cells.html (accessed 15 March 2017)

[86] Fonthal G *et al* 2000 Temperature dependence of the band gap energy of crystalline CdTe *J. Phys. Chem. Sol.* **61** 579–83

[87] Benlattar M, Oualim E M, Harmouchi M, Mouhsen A and Belafhal A 2005 Radiative properties of cadmium telluride thin film as radiative cooling materials *Opt. Commun.* **256** 10–5

[88] http://www.internationalcrystal.net/optics_21.htm (accessed 7 March 2017)

[89] https://www.technologyreview.com/s/600922/first-solars-cells-break-efficiency-record/ (accessed 7 March 2017)

[90] http://topazsolar.com/About-Us/Projects (accessed 7 March 2017)

[91] Del Sordo S, Abbene L, Caroli E, Mancini A M, Zappettini A and Ubertini P 2009 Progress in the development of CdTe and CdZnTe semiconductor radiation detectors for astrophysical and medical applications *Sensors* **9** 3491–526

[92] http://www.eurorad.com/detectors.php#CdTe (accessed 9 March 2017)

[93] https://www.dectris.com/pilatus3_X_CdTe_overview.html#main_head_navigation (accessed 9 March 2017)

[94] Lawson W D, Nielson S, Putley E H and Young A S 1959 Preparation and properties of HgTe and mixed crystals of HgTe-CdTe *J. Phys. Chem. Solids* **9** 325–9

[95] Hansen G L, Schmit J L and Casselman T N 1982 Energy gap versus alloy composition and temperature in HgCdTe *J. Appl. Phys.* **53** 7099

[96] Rogalski A 2005 HgCdTe infrared detector material: history, status and outlook *Rep. Prog. Phys.* **68** 2267–336
[97] Norton P 2002 HgCdTe infrared detectors *Opto-Electron. Rev.* **10** 159–74
[98] Rogalski A 2011 Recent progress in infrared detector technologies *Infrared Phy. Technol.* **54** 136–54
[99] Norton P 2006 Third-generation sensors for night vision *Opto-Electron. Rev.* **14** 1–10
[100] Rogaslki A 2012 History of infrared detectors *Opto-Electron. Rev.* **20** 279–308

Chapter 16

Global infrastructure for emissivity measurements—examples

The need for the acquisition and measurement of repeatable and reproducible data requires the development of standards for measurement techniques with high accuracy. This is particularly challenging for the measurement of the radiative properties of materials. These properties play a vital role in the development of aerospace, commercial and defense industries. World-wide radiative metrology facilities have been created to set methodologies, metrology standards and support the requirements in the study and characterization of a variety of materials such as metals, semiconductors, glasses, plastics and composites. Radiative property measurements are essential for the characterization of optical and thermal properties of materials, as well as in device and systems manufacturing.

During the last decade, significant progress has been made to achieve accuracy of calibration and measurement of spectral emissivity of materials, as a function of their temperature, the emission angle and radiation polarization. These advances are based on the development of optical transmission measurements, detector sensitivity and electronic read-out systems [1].

Emissivity is a radiative property related to the composition and surface morphology of the material. This property characterizes the material capacity to emit electromagnetic radiation, especially in the infrared region of the electromagnetic spectrum [2]. Spectral emissivity, over the thermal infrared spectrum of 1–60 μm, is a decisive optical property that determines the heat energy transfer and balance in the material.

Two important conventional techniques are widely used in the measurement of emissivity. The calorimetric method is the most used technique for the measurement of emissivity of metals, and the other being the Fourier transform spectrometer (FTS) method (radiometric method) [3]. The calorimetric method is a direct heating method that is used for metals and utilizes direct DC current as the heating source; emissivity is determined from the measured surface temperature, area and the electrical power radiating from the surface. The Fourier transform spectrometer method can measure over a wide spectral range with a high spectral resolution and can be used for the study of metals and nonmetals. Both the methods are commonly used but the FTS method allows more accurate emissivity measurements with advanced radiative measurements as a function of material sample angle [4].

16.1 NIST

The National Institute of Standards and Technology (NIST), the oldest physical science laboratory in the United States is part of the US Department of Commerce. NIST provides the measurement standards required by the technology industry, while supporting manufacturing industry competitiveness with an objective to improve the living standards of society [5].

NIST has developed a facility that is capable of characterization of emissivity of materials in the infrared spectral range. This is done by the direct method of radiance in comparison to blackbody reference sources over a temperature range of 200 °C–900 °C, over an extended wavelength range of 1 μm to 100 μm. The method emphasizes the wavelength range of 2–20 μm. Angle-dependent emittance measurements are being implemented, which will also provide sufficient data for the calculation of hemispherical emittance [5].

Figure 16.1 shows the schematic of a state-of-the-art facility designed for the measurement of spectral emissivity of materials. This Fourier transform infrared spectrophotometry (FTIS) facility enables the measurement of spectral directional emittance (emissivity), transmittance and reflectance of materials under the variable control of measurement geometry, beam polarization, and sample temperature [6].

The principle of operation of the facility consists of a set of reference blackbody sources mounted on a motorized stage for selection, interchangeable sample heater/mounts on motorized translation and rotation stages. Additionally, a removable visible/near-infrared integrating sphere for measuring the sample temperature above 500 K is provided. Low scatter interface optics is employed to image the 3 mm to 5 mm central region of the sample or blackbody source that is interfaced on to a water-cooled field stop. The field stop is re-imaged onto either a Fourier transform spectrophotometer equipped with beam-splitters and detectors to cover a spectral range from the visible through the far infrared or a set of filter radiometers mounted on a motorized translation stage for temperature scale transfer between the blackbody sources and for sample temperature determination (together with the integrating sphere). The system further consists of a sectored purge enclosure for the entire beam path, electrical supply, purge gas (Ar, or N_2), and cooling water subsystems, control of system elements and data processing via several PC computers using Lab-View software programs [5].

The sample emissivity is determined through a series of measurement steps. The first step is a measurement of the sample's hemispherical-directional reflectance at the measurement temperature and at a single wavelength matched to the filter radiometer. The second step is the measurement of relative radiance of the sample to blackbody at the same wavelength. The integrating sphere is removed for the second

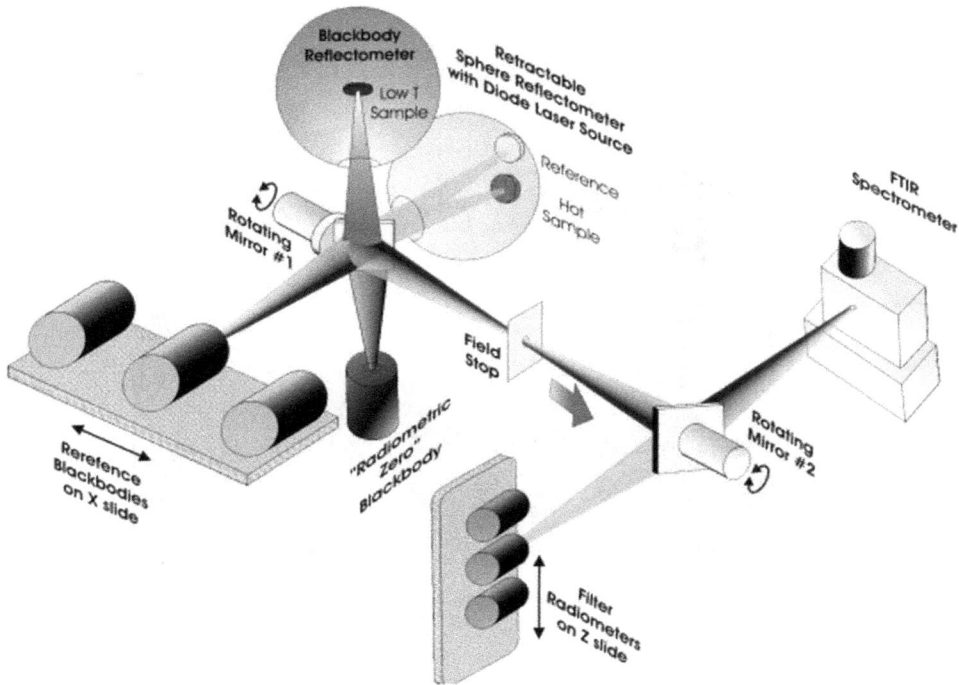

Figure 16.1. Schematic of Infrared Spectral Emittance Characterization Facility [6].

step. The temperature is then calculated from the results of these two steps. In the final step, the Fourier transform infrared spectrometer is used to compare the sample spectral radiance to that of the reference blackbody source as a ratio, and the sample emittance is obtained from the ratio and Planck's law using the sample and blackbody temperatures [5].

16.2 NPL

The National Physical Laboratory (NPL), the largest applied physics organization in the UK is the national measurement standards laboratory for the United Kingdom based at Bushy Park in Teddington, London, England. NPL is an internationally renowned cutting edge metrology center that cooperates with private and government agencies providing measurement standards to scientists and engineers [7]. NPL provides measurement services and manufactures equipment to measure temperature and humidity, force, mass and density, gases and gas analysis, bio-analysis and optical properties among others [8].

NPL supplies and installs a full range of radiometric and photometric measurement solutions and reference calibration standards. NPL has a custom-built instrumentation for optical radiation metrology to customer specific metrology requirements. Standard measurements are made on a high quality commercial

Figure 16.2. Primary Standard Cryogenic Radiometer (NPL) [9].

scanning spectrophotometer whose instrument scales are traceable to national standards. Measurements can be made over the wavelength range from 200 nm to 2500 nm. The bandwidth is typically at 1 nm and reflectance data is commonly reported at 10 nm intervals [7].

NPL's cryogenic radiometer is the primary standard for the measurement of optical radiant power. Figure 16.2 shows a picture of a Standard Cryogenic Radiometer. It uses the electrical substitution technique through which the optical power incident on an absorbing cavity is compared to the electrical power required to heat the cavity to the same temperature. High T_c superconducting leads to the cavity heater ensure true equivalence of electrical and optical power [9]. NPL's cryogenic radiometer allows a laboratory to establish its own scales that are traceable to metrology standards. These scales are usually disseminated using trap detectors as a transfer standard and solid state working standards. The cryogenic radiometer is used with a variety of laser sources.

Table 16.1 shows the specifications for the state of the art Primary Standard Cryogenic Radiometer manufactured by NPL in compliance with established international metrology standards [9].

16.3 Shimadzu

Shimadzu Corp. is a Japanese public company based in Kyoto, Japan, dedicated to manufacturing precision instruments, measuring instruments and medical equipment. Shimadzu Corp. develops and commercializes a broad range of analytical instruments that focus on research, development, and quality control in a wide variety of fields.

Table 16.1. Primary Standard Cryogenic Radiometer specifications [9].

Operating Details	
Cavity operating temperature	12 K
Absolute accuracy	±0.005%
Power Requirements	220 V–240 V, 50 Hz, 1.9 kVA
	208 V–220 V, 60 Hz, 2.0 kVA
Cold head service interval	>15 000 h
Absorber service interval	15 000 h
Water cooling	3 l min^{-1} minimum at 20 °C
Radiometric Properties	
Receiver response	1.2 K mW^{-1} approximately
Maximum power	2 mW
Resolution at maximum power	1 in 105
Time constant	70 s
Brewster window transmission	400 nm to 800 nm 99.96%
Cavity Absorption	200 nm to 200 nm >99.99%
Options	
Vacuum pumping station	
Detector comparator stage	
Laser stabilization facility	
Brewster window assembly optimized for mid infrared use	
Brewster window transmission measurement system	

The detector comparitor stage

This company specializes in manufacturing high precision measurement instruments such as chromatographs, spectrometers, and elemental and surface analysis systems that are essential for product development and quality assurance [10].

Shimadzu's corporate philosophy, 'Contributing to Society through science and technology', is a top management priority that guides its efforts in pursuing the achievement of harmony between global environment preservation and business development. In 1996, Shimadzu Corp. and the United Nations University Institute for the Advanced Study of Sustainability (UNU-IAS) jointly established a capacity-building initiative that provides 10 Asian countries with the analytical scientific knowledge and technology in the efforts to monitor and manage pollutants in the environment as well as develop a global monitoring network in Asia [11].

Shimadzu provides analytical instruments, applications and advisory services related to the operation and maintenance of the instruments. Figure 16.3 presents a photograph of a spectral emissivity measurement system consisting of an FTIR spectrophotometer (IRPrestige-21), a black body furnace, a sample heating

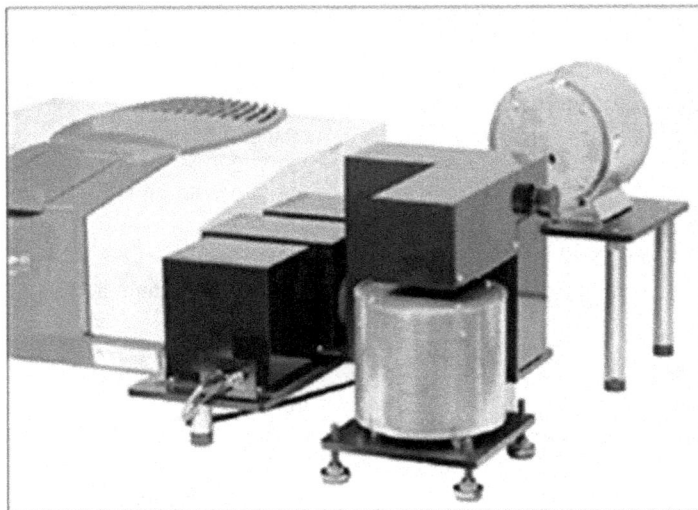

Figure 16.3. Appearance of the Spectral Emissivity Measurement System [12].

furnace, a temperature controller, and a separate optical system. This measurement system uses an FTIR spectrophotometer that complies with the Japanese industrial standard JIS R 1801, and provides data that can be used to evaluate far-infrared radiation materials. JIS R 1801 specifies the method for performing measurement, using an FTIR spectrophotometer, in a wavelength range from 2.5 μm to 25 μm wavelength [12].

16.4 NREL

The National Renewable Energy Laboratory (NREL), located in Golden, Colorado, is the United States' primary laboratory for renewable energy research and development. NREL is funded by United States Department of Energy (DOE) and is a government installation that allows government and private entities to operate the lab. NREL receives funding from the government to be applied toward research, development and support projects, and primarily focus on renewable energy research with emphasis on photovoltaics (PV) under the National Center for Photovoltaics with a number of PV research capabilities including research, testing, development and deployment [13].

NREL has developed an instrument, the PV-Reflectometer for applications in process control and monitoring in the photovoltaic (PV) industry. The PV-Reflectometer is designed for in-line monitoring of several solar cell fabrication steps for all solar cell technologies. This instrument measures the wavelength dependence of reflectance of the entire wafer (or cell), up to 6-in × 6-in in size. The acquired data is analyzed to derive parameters related to surface roughness and texture, antireflection coatings, front-metallization properties (area fraction and thickness), and the back-contact properties of a solar cell. The PV-Reflectometer

Imaging optics

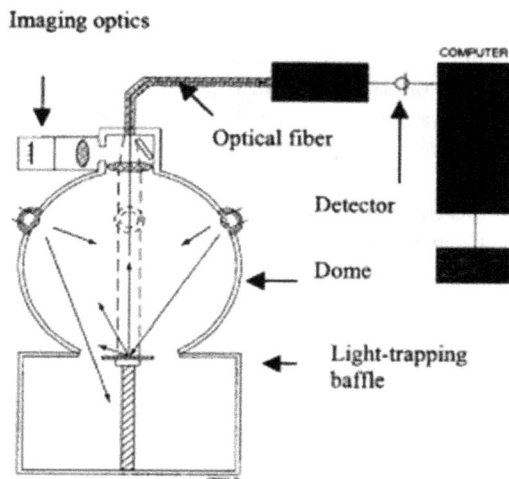

Figure 16.4. A schematic of the PV-Reflectometer [14].

operates according to the concept of 'reciprocal optics', where the sample is illuminated through a very wide angle of incidence and the normal component of the reflected light is measured and analyzed. This apparatus makes the radiation measurement process a rapid and low-cost process. At the same time, it permits the use of high-power sources to enable measurement of 'optically averaged' parameter values for large-area solar cells and wafers. This allows the entire measurement and analysis to be a rapid process that requires less than 10 s, which can be further reduced to less than 1 s [14]. Figure 16.4 shows a schematic of the advanced PV-Reflectometer apparatus developed by NREL.

References

[1] Schmidt V, Meitzner S, Sandring H and King P 2003 Automatic emissivity measurement setup for industrial radiation thermometry *AIP Conf. Proc.* **684** 723–8

[2] Honner M and Honnerová P 2015 Survey of emissivity measurement by radiometric methods *Appl. Opt.* **54** 669–983

[3] Tomáš Králík T, Věra Musilová V, Pavel Hanzelka P and Frolec J 2016 Method for measurement of emissivity and absorptivity of highly reflective surfaces from 20 K to room temperatures *Metrologia* **53** 743–53

[4] Zhang B, Redgrove J and Clark J 2004 *A Transient Method For Total Emissivity Determination* (New York: Plenum) 423–38

[5] https://www.nist.gov/ (accessed 10 February 2017)

[6] Hanssen L, Mekhontsev S and Khromchenko V 2004 Infrared spectral emissivity characterization facility at NIST *Thermosense* XXVI; Burleigh D D, Cramer K Elliott and Peacock G Raymond *Proc. SPIE* **5405** 1–12

[7] http://www.npl.co.uk (accessed 10 February 2017)

[8] http://www.npl.co.uk/measurement-services/ (accessed 10 February 2017)

[9] http://www.npl.co.uk/optical-radiation-photonics/radiometry-+-detectors/products-and-services/primary-standard-cryogenic-radiometer (accessed 10 February 2017)

[10] http://www.shimadzu.com (accessed 10 February 2017)

[11] http://www.shimadzu.com/an/unu-ias_project/overview.html (accessed 11 February 2017)

[12] http://www.shimadzu.com/an/ftir/support/tips/letter13/emissivity.html (accessed 12 February 2017)

[13] http://www.nrel.gov/ (accessed 12 February 2017)

[14] Sopori B, Zhang Y, Chen W and Madjdpour J 2000 Silicon solar cell process monitoring by PV-reflectometer *IEEE* 120–3

www.ingramcontent.com/pod-product-compliance
Lightning Source LLC
Chambersburg PA
CBHW081543220326
41598CB00036B/6541